PROGRESS IN BIOMETEOROLOGY

EDITOR-IN-CHIEF: S. W. TROMP MANAGING EDITOR: JANNEKE J. BOUMA

DIVISION A

PROGRESS IN HUMAN BIOMETEOROLOGY

The Effect of Weather and Climate on Man
and his Living Environment

VOLUME 1 PART III PERIOD 1963 - 1970

BIOMETEOROLOGICAL ASPECTS OF PLANTS, TREES AND ANIMALS IN HUMAN LIFE

EDITED BY S. W. TROMP
BIOMETEOROLOGICAL RESEARCH CENTRE, LEIDEN
THE NETHERLANDS
IN COOPERATION WITH 90 CONTRIBUTORS

SZ

SWETS & ZEITLINGER N.V. AMSTERDAM 1972

Printed in photo-litho-offset by Drukkerij Miedema & Co., Leeuwarden, The Netherlands

ISBN 90 265 0156 0

LIST OF CONTRIBUTORS OF VOLUME 1

NAME	ADDRESS	PART CHAPTER SECTION
Dr Elisabeth ALBRECHT	Mehringerstrasse 14, 1 Berlin 28, GERMANY	II,7.12*
Dr H. ALBRECHT	Mehringerstrasse 14, 1 Berlin 28, GERMANY	II,7.12
Dr J. AUBERT	Clinique Pneumo-Phtisiologique, Hôpital Ste Marguérite, Chemin de Ste Marguérite, Marseille 9, FRANCE	I,2.1
Dr W. BAIER	Agrometeorology Section, Plant Research Inst., Central Experimental Farm, Ottawa, CANADA	III,9.0
Dr L. BAILLAUD	Lab. de Biologie Végétale, Fac. des Sciences, 4 Rue Ledru, F–63 Clermont-Ferrand, FRANCE	I,6.21c
Prof. Madeleine F. BARNOTHY	Dept. of Physics, College of Pharmacy, Univ. of Illinois, Chicago, Ill., USA	1,6.14
Dr W. BIANCA	Inst. f. Tierzucht ETH, Universitätstr. 2, Zürich 8006, SWITZERLAND	I,6.6a
Dr K. BIERSTEKER	Gem. Geneeskundige en Gezondheidsdienst, Centr. Gebouw v. Volksgezondheid, Baan 170, Rotterdam, THE NETHERLANDS	I,2.3b
Dr J. BLIGH	Agricultural Research Council, Inst. of Animal Physiology, Babraham, Cambridge, GREAT BRITAIN	I,6.17a
Miss Janneke J. BOUMA	Biometeorological Research Centre, Leiden Haarlemmertrekvaart 33, Oegstgeest (Leiden), THE NETHERLANDS	II,7.2, 7.5b (2), 7.22b (1)
Dr C. BOUTIN	Clinique Pneumo-Phtisiologique, Hôpital Ste Marguérite, Chemin de Ste Marguérite, Marseille 9, FRANCE	I,2.1

* II,7.12 means Part II, Chapter 7, Section 12.

Mr D. BRANAGAN	Centre f. Tropical Veterinary Medicine, Univ. of Edinburgh, Dept of Animal Health, Royal (Dick) School of Veterinary Studies, Veterinary Field Station, Easter Bush, Roslin, Midlothian, Scotland, GREAT BRITAIN	III,10.6
Mr J.R. BRECKENRIDGE	U.S.Army Research Inst. of Experimental Med., Natick, Mass. 01760 U.S.A.	
Dr H. BRENN	Gäuggelistr. 6, CH 7000 Chur, SWITZERLAND	II,7.10
Prof. J.C. BROCKLEHURST	Dept of Geriatric Medicine, Univ. of Manchester, Univ. Hospital of S. Manchester, Nell Lane, Manchester M20-8LR, GREAT BRITAIN	II,7.14d
Prof. K.W. BRÜCK	Dept of Physiology II, Justus Liebig Univ., Friedrichstr. 24, 6300 Giessen, GERMANY	I,6.17b(1)
Prof. G.E. BURCH	Dept of Medicine, Tulane Univ., 1430 Tulane Ave, New Orleans, La. 70112, USA	II,7.6
Ing. Carmen CAPEL-BOUTE	C.I.R.E.F.A., Univ. Libre de Bruxelles, 50 Av. F.D. Roosevelt, Brussels 5, BELGIUM	I,3.1
Dr A.P. CASARETT	Dept of Physical Biology, N.Y. State Veterinary College, Cornell Univ., Ithaca, N.Y. 14850, USA	I,6.11
Dr A. CHAN	Plant Research Inst., Central Experimental Farm, Ottawa, CANADA	III,9.0
Dr J. CHARPIN	Clinique Pneumo-Phtisiologique, Hôpital Ste Marguérite, Chemin de Ste Marguérite, Marseille 9, FRANCE	I,2.1
Dr Marjorie CLIFTON	6 Cedar Drive, Keyworth, Notts. NG12-5AB, GREAT BRITAIN	II,7.3
Prof. J.L. CLOUDSLEY-THOMPSON	Dept of Zool., Birkbeck College, Malet Str., London WC 1-E7HX, GREAT BRITAIN	I,6.21b, 6.22b (1)
Dr R. DAVIES	The Wright-Fleming Inst. of Microbiology, St. Mary's Hospital Medical School. Allergy Dept, Norfolk Place, London W2, GREAT BRITAIN	II,7.1a

Dr T.R.A. DAVIS	Arthur D. Little Inc., Life Sciences Div., Cambridge, Mass., 02140, USA	I,6.2, 6.3
Prof. B. DIECKHUES	Univ. Augenklinik, Westring 15, 44 Münster, GERMANY	II,7.5b (1)
Dr K. DIRNAGL	Balneologisches Inst., Med. u. Klim. Abt., Marchioninistr. 17. 8 München 55 (Groszbadern), GERMANY	I,2.2, 6.10a II,7.22a (2)
Dr J.M. DONEY	Hill Farming Research Organization, 29 Lauder Rd, Edinburgh 9, Scotland, GREAT BRITAIN	III,10.2a
Prof. D.F. DOWLING	Dept of Animal Husbrandry, Univ. of Queensland, St Lucia, AUSTRALIA	III,10.1b
Prof. C.J. EAGAN	Dept of Physiology and Biophysics, Colorado State Univ., Fort Collins, Col. 80521, USA	II,8.11a
Dr Marjorie EDMAN	Dept of Nutrition and Biochemistry, Animal Science Div., College of Agriculture, Univ. of Illinois, Urbana, Ill.61801, USA	II,7.20
Surgeon-Captain F.P. ELLIS	Formerly, Royal Naval Medical School, Alverstoke, Gosport, Hamps., GREAT BRITAIN. Present address: 27 Belsize Court, Wedderburn Rd, Hampstead, London N.W.3, GREAT BRITAIN	I,4.4
Dr P.O. FANGER	Lab. for Varme-og Klimateknik, Danmarks Tekniske Høјskole, Bygning 402, DK 2800 Lyngby, DENMARK	I,4.1
Mr W. FERGUSON	Organización de las Naciones Unidas para la Agricultura y la Alimentación, Officina Regional para América Latina, Avd. Providencia 871, Casilla 10095, Santiago de Chile, CHILE	III,10.6
Prof. G.E. FOLK Jr	Dept of Physiol., State Univ. of Iowa, Iowa City, Iowa, USA	I,6.22b (2)
Dr K.I. FURMAN	Dept. of Physiol., Med. School, Univ. of Witwatersrand, Hospital Str., Johannesburg, S. AFRICA	II,7.17
Dr H. GEHRKEN	Klinik f. Dermatologie und Allergie, Hochgebirgs-Sanatorium Valbella, Davos-Dorf, CH 7260, SWITZERLAND	II,7.10

Dr T.D. GILES	Dept. of Medicine, Tulane Univ., 1430 Tulane Ave, New Orleans, La. 70112, USA	II,7.6
Dr B. GIVONI	Dept. of Building Climatology, Building Res. Station, Technion, Haifa, ISRAEL	I,4.2, 4.3, 5.0
Prof. J. GOLD	Syracuse Cancer Inst. Inc., Presidential Plaza, 600 E. Geneseestr., Syracuse, N.Y. 13202, USA	II,7.13a
Dr R.F. GOLDMAN	Director military Ergonomics Lab., U.S.Army Research Inst. of Experimental Med., Natick, Mass. 01760, USA	II,7.19
Dr H. de GRAAF	Former Head Lab. for Air Pollution Research Rotterdam, Karmelweg 60, Rotterdam, THE NETHERLANDS	I,2.3a
Prof. F. HALBERG	Dept. of Pathology, Univ. of Minnesota, Health Sciences Center, Minneapolis, Minn. 55455, USA	I,6.21d
Dr E. HAMMELBURG	Ear-Nose-Throat Dept, Leyenburg Hospital, Leyweg, The Hague, THE NETHERLANDS	I,6.16
Dr J.B. HANSEN	Dept. of Thoracic and Cardiovascular Surgery, Odense Univ. Hospital, Kløvervaenget, Odense, DENMARK	II,7.11
Dr W.O. HAUFE	Vet. and Med. Entomology Section, Canada Agricultural Research Station, Lethbridge, Alb., CANADA	III,10.7
Prof. H. HENSEL	Physiol. Inst. d. Univ., Deutschhausstr. 2, 355 Marburg/Lahn, GERMANY	I,6.17a
Prof. F. HOLLWICH	Univ. Augenklinik, Westring 15, 44 Münster, Westfalen, GERMANY	I,6.12b (1), 7.5b (1)
Mr G.W. HURST (died Jan. 1971)	Formerly Meteorological Office, Bracknell, Berks., GREAT BRITAIN	II,7.7b
Dr J.C.D. HUTCHINSON	Div. of Animal Physiol. Ian Clunies Ross Animal Research Lab.-Prospect, CSIRO, P.O.B. 239, Blacktown, N.S.W.2148, AUSTRALIA	III,10.2b
Dr D.L. INGRAM	Agricultural Research Council, Inst. of Animal Physiol., Babraham, Cambridge, GREAT BRITAIN	III,10.3

Prof. H.D. JOHNSON	Dept. of Dairy Husbandry, College of Agriculture, 208 Eckles Hall, Univ. of Missouri, Columbia, Mo., USA	III,10.1a
Prof. E. JOKL	Research Lab., Univ. of Kentucky, Lexington, Kent. 40506, USA	II,8.11b
Dr N.E. JONASSEN	Lab. of Applied Physics, Techn. Univ. of Denmark, Bldg 307, DK 2800 Lyngby, DENMARK	I,4.1
Dr G.H. JONKERS	Burg. Knappertlaan 92, Schiedam, THE NETHERLANDS	II,7.5a
Prof. H. JUNGMANN	II Medizinische Universitätsklinik und Poliklinik, Martinistr. 52, 2 Hamburg-Eppendorf, GERMANY	I,6.6b, 6.19b
Prof. H.J. JUSATZ	Inst. f. Tropenhygiene u. öffentl. Gesundheitswesen, Süd Asien Inst. der Univ., Neue Schlossstr. 3, 89 Heidelberg, GERMANY	II,7.18
Dr M.B. KREIDER	Biology Dept, Worcester State College, Worcester, Mass. 01602, USA	II,7.13b
Dr E. KRIEGER	Klinisches Sanatorium Trausnitz, Salzburgerstr. 9, 8230 Bad Reichenhall, GERMANY	I,6.10a, 7.22a (2)
Prof. A.P. KRUEGER	School of Publ. Health, Earl Warren Hall, Univ. of California, Berkeley, Cal. 94720, USA	I,6.9
Prof. H.E. LANDSBERG	Inst. f. Fluid Dynamics and Applied Mathematics, Univ. of Maryland, College Park, Md. 28742, USA	I,1.2
Dr J.S. LAWRENCE	The Arthritis and Rheumatism Council, 128 Birchfields Rd, Manchester M14-6PF, GREAT BRITAIN	II,7.9
Prof. W.V. MACFARLANE	Dept. of Physiology, Waite Agricult. Res. Inst., Univ. of Adelaide, Private Bag, 1, Glen Osmond, Adelaide (S.A. 5064), AUSTRALIA	I,6.18, 6.19a, 6.22a (4)
Dr B.S. MACKIE	Macquerie Chambers, 183 Macquerie Str. Sydney, AUSTRALIA	II,7.4b
Prof. N.G. MEYNE	Dept. of Cardio-Vascular Surgery, Wilhelmina Gasthuis, 1e Helmerstr. 104, Amsterdam, THE NETHERLANDS	II,7.22b (2)
Prof. R. MILINE	Inst. d'Histologie et d'Embryologie, Fac. de Médécine, Hajduk Veljkova, Novi Sad 7-9, YUGOSLAVIA	I,6.12a

Dr Masako MOMIYAMA	Meteorological Research Inst., Koenji Kita 4-35-8, Suginami-Ku, Tokyo, JAPAN	I,6.22a (2)
Dr T. MORIMOTO	Dept. of Physiology, Univ. of Medicine, Kawara-Machi, Kyoto, JAPAN	I,6.22a (3)
Prof. P.R. MORRISON	Zoophysiological Lab., Univ. of Alaska, Fairbanks, Alaska, USA	I,6.5
Prof. E.A. PETROPOULOS	Dept. of Anatomy, School of Medicine, Univ. of Calif., San Francisco, Cal. 94122, USA	I,6.7
Prof. G. PICCARDI	Centro Universitario Fenomeni Fluttuanti, 26 Via Gino Capponi, I 50121, Florence, ITALY	I,3.1
Dr Z. PUCEK	Mammals Research Inst., Polish Acad. of Sciences, Bialowicza, Woj. Bialystok, POLAND	II,8.6
Prof. A. REINBERG	Fondation A. de Rothschild, Lab. de Physiologie, 29 Rue Manin, 75 Paris 19, FRANCE	I,6.4b
Dr R. REITER	Physikalisch-Bioklimatische Forschungsstelle, Skistadion, Westtor, 81 Garmisch- Partenkirchen, Bayern, GERMANY	I,1.1, 6.8
Dr J.F. SAUNDERS	Chief Biology Program, Bioresearch Div., Office of Life Sciences, National Aeronautics and Space Adm., Washington D.C. 20546, USA	I,3.2, 6.13, 6.15, 6.19c
Prof. W. SCHMIDT-KESSEN	Inst. f. Balneologie u. Med. Klimatologie, Univ. Freiburg, Herderstr., Freiburg/Br., GERMANY	II,7.22c, 7.22d
Dr Heide SCHMIEGER	Univ. Augenklinik, Westring 15, 44 Münster, Westfalen, GERMANY	I,6.12b (2)
Dr J. SLEE	Animal Breeding Research Org., Edinburgh 9, Scotland, GREAT BRITAIN	III,10.2a
Prof. D. SMIDT	Inst. f. Tierzucht u. Haustiergenetik, Thaerweg 1, 34 Göttingen, GERMANY	III,10.5
Prof. A. SOLLBERGER	Depts of Physiology and Information Proc., School of Medicine, Southern Illinois Univ., Life Science 1, Carbondale, Ill. 62901, USA	I,6.21a
Prof. W.H. TERJUNG	Dept. of Geography, Univ. of California, Los Angeles, Cal. USA	I,1.3, 4.5

Dr J.C. THOMPSON	Dept. of Physical Biology, New York State Veterinary College, Cornell Univ., Ithaca, N.Y. 14850, USA	I,6.11
Prof. Paola S. TIMIRAS	White Mountain Research Station, Univ. of Calif., Berkeley, Cal. 94720, USA	I,6.7
Dr S.W. TROMP	Biometeorological Research Centre, Leiden, Haarlemmertrekvaart 33, Oegstgeest (Leiden), THE NETHERLANDS*	*
Dr P.L. VON DESCHWANDEN Jr	Klimaforschungsstation, CH 3715 Adelboden, SWITZERLAND	II,7.14c
Dr A.P. WEHNER	Battelle Memorial Inst., P.O.B. 999, Richland, Wash. 99352, USA	I,6.10b, II,7.22a (1)
Dr W.H. WEIHE	Biologisches Zentrallaboratorium, Kantonspital Zürich, Univ. Zürich, Rämistr. 100, 8006 Zürich, SWITZERLAND	III,10.8
Prof. W.O. WILSON	Dept. of Poultry Husbandry, Agric. Exp. Station, College of Agriculture, Univ. of Calif., Davis, Cal. 95616, USA	III,10.4
Dr P. WISE	Endocrine Unit, Royal Adelaide Hospital, North Terrace, Adelaide, S.A. 5000, AUSTRALIA	II,7.14b
Prof. H. YOSHIMURA	Formerly, Dept. of Physiol., Univ. of Medicine, Kawara-Machi, Kyoto, JAPAN	I,6.22a (3)

* Sections covered by Dr S.W. Tromp:
I,6.1, 6.4a, 6.17b (2), 6.22a (1); II,7.1b, 7.1c, 7.2, 7.4a, 7.5b (2), 7.7a, 7.8, 7.14a, 7.14e, 7.15, 7.16, 7.21, 7.22b (1), 8.1, 8.2, 8.3, 8.4, 8.5, 8.7, 8.8, 8.9, 8.10.

PART III

BIOMETEOROLOGICAL ASPECTS OF PLANTS, TREES AND ANIMALS IN HUMAN LIFE

PREFACE

Since 1963 Biometeorology has been developing rapidly due to the considerable increase in interest of the whole scientific world in man's physical environment. Also the recent developments in Space Research in connection with the landings on the moon and studies of the atmosphere of the planets have stimulated many research workers to enter this field of research.

The present book "PROGRESS IN HUMAN BIOMETEOROLOGY" is a continuation of the textbook "MEDICAL BIOMETEOROLOGY" (Weather, Climate and the Living Organism) which was prepared by the Editor in 1962 in close cooperation with 26 contributors and published by Elsevier Publishing Company, Amsterdam in 1963.

Swets & Zeitlinger Publishing Company agreed to publish a continuation of "Medical Biometeorology", starting with the period 1963–1970. As the effects both on healthy man and on diseased subjects will be reviewed, the title was changed to "PROGRESS IN HUMAN BIOMETEOROLOGY" with the subtitle "THE EFFECTS OF WEATHER AND CLIMATE ON MAN AND HIS LIVING ENVIRONMENT".
It was felt that the meteorological effects on living organisms in the immediate environment of man, which are of great importance for human life, should also be included in the present volumes, although part of these phenomena affect the human body only indirectly. This explains the inclusion of two chapters, covering the significance of plant and animal life for man.

However, not only "Human Biometeorology" needed a new up to date textbook, but also the recent developments in plant and animal biometeorology made it necessary to cover these developments. For this reason this new series will be called PROGRESS IN BIOMETEOROLOGY. It will consist of three divisions: Division A: PROGRESS IN HUMAN BIOMETEOROLOGY; Division B: PROGRESS IN ANIMAL BIOMETEOROLOGY and Division C: PROGRESS IN PLANT BIOMETEOROLOGY. Each division will have its own Editor. Division A is being prepared by Dr. S.W. TROMP, Director of the Biometeorological Research Centre, Leiden, The Netherlands; Division B will be prepared by Prof. H.D. JOHNSON, Director of the Department of Dairy Husbandry, University of Missouri, Columbia, USA; Division C by Dr. L.P. SMITH, former President of the Commission for Agricultural Meteorology of the World Meteorological Organization, Meteorological Office, Bracknell, Great Britain. General Coordinator and Editor-in-Chief of the three divisions is Dr. S.W. TROMP. Managing Editor of all volumes is Miss JANNEKE J. BOUMA, Research Associate of the Biometeorological Research Centre, Leiden, The Netherlands. Depending on the amount of new information received by the

Editors in their specific divisions, new volumes will be prepared every two or three years. For this reason the present book is called DIVISION A, VOLUME 1 (PERIOD 1963–1970).

During the preparation of Division A Volume 1 it became evident that the coverage of all biometeorological developments since 1963 would require such a large volume that it would be difficult to handle this in one book. For this reason the Editor decided to split volume I into three parts:

Part I covers THE MICRO- AND MACRO-ENVIRONMENTS IN THE ATMOSPHERE AND THEIR EFFECTS ON BASIC PHYSIOLOGICAL MECHANISMS OF MAN

Part II covers PATHOLOGICAL BIOMETEOROLOGY, in other words the effects of weather and climate on various diseases and the applications of climatotherapy for the cure or improvement of these diseases.

Part III covers BIOMETEOROLOGICAL ASPECTS OF PLANTS, TREES AND ANIMALS IN HUMAN LIFE and consists of two chapters only:
Chapter 9: DIRECT EFFECTS OF THE PHYSICO-CHEMICAL ENVIRONMENT OF PLANTS AND TREES ON HUMAN LIFE
Chapter 10: METEOROLOGICAL EFFECTS ON ANIMALS IMPORTANT FOR MAN, in which the effects of Cattle, Sheep, Pigs, Poultry, Fertility and Animal Reproduction, Animal Diseases, Arthropods (as vectors of Human and Animal Disease) and Laboratory Animals have been reviewed by different experts in these specialized fields.

The detailed contents of the three Parts of Volume 1 will follow this Preface. The list of 90 contributors of Division A, Volume 1 with their addresses and Sections covered by each of them is given on page 1-7.

Instead of giving a list of references after each Chapter or Section we follow the method used in "MEDICAL BIOMETEOROLOGY", which proved to be very satisfactory. All references are given at the end of each book arranged in the same sequence as the Chapters and Sections. At the end of each Section we briefly indicated the pages where the references of this particular Section can be found. An extensive Author and Subject Index will enable the reader to find each Section in the book and the Contributors of these Sections.

This Volume 1 represents in a way the standard type of Progress Series but differs from the usual publications in which a number of authors review only certain aspects of a whole science. In this Progress Series we have tried to prepare up to date textbooks on Human Biometeorology which, together with "Medical Biometeorology", cover the whole period from 1900 onwards with a list of thousands of references. In two or three years time the next volume will keep these textbooks up to date.

The same principles described for Division A, Volume I and following volumes will be applied in Division B (Animal Biometeorology) and Division C (Plant Biometeorology).

We fully realize that the task ahead of us is a very difficult one, but we trust that the efforts of all these experts in the field of Biometeorology will enable us to keep abreast with the fast developments of this modern branch of science and will help future students in Meteorology, Biology, Medicine, Veterinary Science and many other sciences, in finding their way in this dense forest of rapidly growing scientific knowledge.

S. W. Tromp

GENERAL CONTENTS OF VOLUME 1, Parts I, II and III

PART I

MICRO- AND MACRO-ENVIRONMENTS IN THE ATMOSPHERE THEIR EFFECTS ON BASIC PHYSIOLOGICAL MECHANISMS OF MAN

PART II

PATHOLOGICAL BIOMETEOROLOGY

PART III

BIOMETEOROLOGICAL ASPECTS OF PLANTS, TREES AND ANIMALS IN HUMAN LIFE

INTRODUCTION

Contrary to our original planning Part III, Volume I, Division A, in stead of Part I, is the first book which appears in the series "PROGRESS IN HUMAN BIOMETEOROLOGY". Various, mainly organizational reasons forced us to publish first Part III before Parts I and II.
One of the reasons is that Part III is a subject in itself, slightly different from Parts I and II and readers specially interested in these topics may be less interested in Parts I and II, which they may order at a later date.

Secondly, it is the least complicated part of these three textbooks and could be completed therefore at an earlier date than Parts I and II. This is particularly true as some of the 90 contributors of very important sections of Part I could not keep their promises due to serious illness. The time which elapsed in finding new contributors and preparation of their manuscripts made it impossible to print Part I before Part III. However, the Editors hope that Part I will appear only a few months after the publication of Part III.

The Editor-in-Chief wishes to avail himself of this opportunity of thanking all Contributors for their most valuable assistance and wholehearted cooperation and Swets & Zeitlinger Publishing Company in particular for their courage to publish this new Series in the field of Biometeorology. If these publications may help to solve the most urgent problems of the increasingly polluted environment, the Contributors will be all amply rewarded for the many working hours they have sacrificed to fulfil their difficult task.

S.W. Tromp

DEFINITION AND CLASSIFICATION OF BIOMETEOROLOGY

Various definitions and classifications have been suggested since the Foundation of the International Society of Biometeorology (I.S.B.) in 1956. Biometeorology can be defined briefly as the study of the biological effects of weather and climate on living organisms, plants, animals and man and of their direct physico-chemical environment.

In 1956, during the first Biometeorological Symposium of the International Society of Biometeorology, at UNESCO Headquarters in Paris, the following definition was accepted by the members of ISB:

"Biometeorology comprises the study of the direct and indirect interrelations between the geophysical and geochemical environment of the atmosphere and living organisms, plants, animals and man. The term "environment" is broadly conceived and includes micro, macro and cosmic environments in so far as they affect the earth's atmosphere and the diverse physical and chemical factors which comprise these environments. The investigations in these disciplines should be conducted in nature or in the laboratory under as rigidly controlled conditions as possible to describe measurable and reproducible physical, chemical and biological factors which show a sufficiently high statistical correlation with measurable physiological and pathological processes to suggest a valid cause and effect relationship between organism and environment."

Since 1969 the members of ISB were not satisfied anymore with this very long definition and the greatest majority of the members from 53 countries decided to use for official purposes the short definition only, as indicated above.

However, this short definition is insufficient if the entire field of biometeorological activities should be sharply defined. For this reason another slightly longer definition was suggested indicating the various types of phenomena studied by biometeorology. According to this definition biometeorology comprises the study of the direct and indirect effects (of an irregular, fluctuating or rhythmic nature) of the physical, chemical and physico-chemical micro- and macro-environments of both the earth atmosphere, and of similar extra-terrestrial environments, on physico-chemical systems in general and on living organisms (plants, animals and man) in particular.

The biometeorological sciences have been divided into six main groups: phytological, zoological, human, cosmic, space and paleo-biometeorology.

1. PHYTOLOGICAL BIOMETEOROLOGY studies the influence of weather and climate on the development and distribution of plants for general phytological, agricultural and forestry purposes, and also the effects on small plant organisms responsible for the development of piant, animal and human diseases.

2. ZOOLOGICAL BIOMETEOROLOGY studies the influence of weather and climate on animals in general, on milk production, reproduction and coating of cattle, etc. *(Veterinary biometeorology)*, on egg-laying rhythm, etc., of birds *(Avian biometeorology)* but in particular the effect of weather and climate on insects and terrestrial arthropoda, their physiological activities and development and the outbreak of plant, animal and human diseases as a result of these developments, such as the abundant development of locusts, malaria mosquitoes, etc. *(Entomological biometeorology)*.
3. HUMAN BIOMETEOROLOGY studies the influence of weather and climate on man. It is usually divided into five important sections:
   a. PHYSIOLOGICAL BIOMETEOROLOGY: the influence of weather and climate on physiological processes in normal healthy man;
   b. SOCIAL BIOMETEOROLOGY: use of favorable meteorological factors for the treatment of large population groups as preventive or curative measures;
   c. PATHOLOGICAL BIOMETEOROLOGY: the influence of weather and climate on the various physiological and pathological phenomena associated with the diseases of man, the period of outbreak, intensity and geographical distribution;
   d. ARCHITECTURAL AND URBAN BIOMETEOROLOGY: the influence of the microclimates in houses and cities on the health of man and the effect of architectural constructions and town planning on these microclimates;
   e. NAUTICAL BIOMETEOROLOGY: the influence of weather and climate on animals and man living on moving ships.

   Of special importance for the general practioner are the results obtained in *Acclimatization biometeorology* (a branch of physiological biometeorology which studies the various physiological processes involved during adaptation to extreme conditions of temperature, humidity, etc., e.g., in the tropics or polar regions); *Air pollution biometeorology* (influence of pollen, spores, dust and chemical air pollution on man) and the various methods applied in Climatotherapy.

   In Table 1 a more detailed classification of Human Biometeorology is given.
4. COSMIC BIOMETEOROLOGY studies the influence of extraterrestrial factors such as variations in solar activity, cosmic radiation, etc., on living organisms on earth.
5. SPACE BIOMETEOROLOGY is closely related to space research. It studies the microclimatic conditions in space vehicles and the biological effects (both observed and expected) of extra terrestrial physical environments on man and animals on the moon and the planets.
6. PALEO-BIOMETEOROLOGY studies the influence of climatic conditions in the past (millions of years ago) on the evolutional development and geographic distribution of plants, animals and man.

Table 1. Classification of HUMAN BIOMETEOROLOGY

1. PHYSIOLOGICAL BIOMETEOROLOGY
   a. General physiological biometeorology
   b. Geographical biometeorology
   c. Ethnological biometeorology
   d. Acclimatization biometeorology
2. SOCIAL BIOMETEOROLOGY
   a. General social biometeorology
   b. Psychological biometeorology (including aestheto-biometeorology)
   c. Archaelogical biometeorology
3. PATHOLOGICAL BIOMETEOROLOGY
   a. General pathological biometeorology
   b. Meteorological pathology
   c. Climatological pathology
   d. Air pollution biometeorology
      (1) Pollution with organic particles (pollen, spores, etc.)
      (2) Pollution with inorganic particles (dust, etc.)
      (3) Chemical pollution
      (4) Aerosol biometeorology
   e. Geographical climatopathology
   f. Climatotherapy (or therapeutic biometeorology)
      (1) General climatotherapy
      (2) Thalassotherapy (climatological)
      (3) Heliotherapy
      (4) Thermotherapy
      (5) Aerosol (and ionization) therapy
      (6) Socio-climatotherapy
      (7) Climatic health resorts
      (8) Climatic chamber treatments
   g. Pharmacological biometeorology
4. ARCHITECTURAL AND URBAN BIOMETEOROLOGY
   a. General urban biometeorology
   b. Architectural biometeorology
   c. Sanatorium biometeorology
5. NAUTICAL BIOMETEOROLOGY
   a. General nautical biometeorology
   b. Cargo biometeorology

S.W. TROMP

PART III

CHAPTER 9

DIRECT EFFECTS OF THE PHYSICO-CHEMICAL ENVIRONMENT OF PLANTS AND TREES ON HUMAN LIFE

Direct Effects of the Physico-Chemical Environment of Plants and Trees on Human Life

by

W. Baier and A. Chan

1. INTRODUCTION

Plants in the broadest sense from the basis for making life on the planet Earth possible. In this review we shall be dealing for the most part with the effects of terrestrialplants but it should be understood that marine plants play an enormous role in the physical and chemical environment of the world. For example, 88 per cent of photosynthesis takes place in the oceans as compared with 7.9 per cent in forests and 3.1 per cent in cultivated lands (Spector, 1956). It can be readily seen that the plants of the ocean have a tremendous influence in both carbon dioxide and oxygen cycles. Any changes that do occur are of such a time scale that they cannot be termed direct and the implications will not be discussed here.

In the space available for this Chapter, we have selected areas which seem to be the most pertinent in the present content of dealing with progress in human bio-meteorology. The important role of plants for human confort, health and survival has recently received new impetus when the dramatic consequences of man's changing the environment became so alarming. It is, therefore, appropriate to concentrate on plant aspects related to man's efforts in controlling pollution of the environment. This is like looking in the reverse at the process of pollution of the environment by man, a subject of much public concern and controversy. Plants act as air pollution detectors and dust traps, reducers of noises in cities, factories and airports, and they create a micro-environment to enhance our surroundings.

Numerous references to the problem of environmental pollution are scattered over a variety of scientific and miscellaneous publications. But reported experiments and data on the role of plants as air conditioners or air purifiers are scarce, because research in this field has just begun. Because of this, it should be recognized that the present review is incomplete and can only cite typical examples from the international literature.

Although the aesthetic aspects of planted or natural vegetation is only dealt with briefly, the potential solutions to some environment crises offered by plants have not yet been fully realized. The scientific data on environmental improvement through landscape design is insufficient to be useful. However, the situation is being corrected. The American Society of Landscape Architects Foundation, founded by the U.S. National Park Service and the National Endowment for the Arts, is compiling an illustrated list of the possible functions of plants in modern society. Although the list is not yet complete, it includes erosion control, acoustical control, atmospheric purification, traffic control, as well as aesthetic uses.

2. RELATIONSHIPS BETWEEN PLANTS AND AIR POLLUTION

The effects of pollutants on plants, as on animals and man, are as varied as the different air pollutants, the recipient organism, and their various interactions with other factors in the environment. Recently, Foy (1970) aptly presented the relationships between plants and pollution. "Plants bear an important relationship to the subject of pollution in at least four ways: (1) Plants sometimes pollute the environment; (2) Plants may act as carriers of pollutants; (3) Plants are injured by various environmental pollutants which reduce the quantity or quality of plant products; and (4) Plants may remove pollutants directly or otherwise serve to mitigate the adverse effects of pollution in certain segments of the biosphere." Fortunately, the contribution by plants to atmospheric pollution is insignificant as compared with their beneficial role as air pollution detectors, dust traps, and air pollutant removers. Examples for these various roles of plants follow.

PLANTS AS AIR POLLUTERS

Foy (1970) refers to phototoxic gaseous emanations, leachates, and root exudates from certain trees and shrubs inhibiting the growth of surrounding vegetation. But of special interest here is the work by Went (1966). He concluded from his research into the effect of smog on plants and animals that not only man-produced pollution products, such as gasoline vapors and exhaust gases from engines, brought about air pollution but also volatile materials produced by vegetation can do the same thing. Since approximately one thousand million tons of volatile organic substances are released per year by the vegetation over the whole world, one can imagine that this presents a special problem as far as its disposal is concerned. He found that anywhere in the world and at anytime that analyses were carried out, a measurable amount of organic substances, tentatively identified as TERPENES, were in the air at a concentration of 2 to over 20 parts per thousand million. The same general composition of the air was found in which plants had been enclosed. There was a gradient from within the vegetation to the air above it of these compounds. This indicates that plants are the source of these terpenes. Such volatile materials produced by plants may have a considerable effect on other plants. Went (1966) suggested that the dissipation of terpenes and other volatile plant products passes through the same cycle as the dissipation of gasoline vapors in cities. This means that after their release by plants they are photochemically altered by sunlight. This results first in the formation of peroxides, ozonides, or free radicals which later leads to the production of large quantities of so-called "Aitken condensation nuclei".

Went (1960) also suggested that volatilization of terpenes and other plant products results in the production of, first, blue haze, then veil clouds, and finally, bituminous materials which, when returned to the earth by rain and snow, produce the source materials for petroleum formation. The blue hazes and veil clouds play very important parts in the heat balance of the earth and especially of the atmosphere. They probably absorb about 20 per cent of all the sun's radiation reaching the earth. Therefore, they should be considered when calculating the heat conversions near the earth's surface. He presents the hypothesis that trees serve as a source of atmospheric toxicants because their emmitted aromatic hydrocarbons can become involved in photochemical reactions with the oxygen and oxides of nitrogen in the air, thus producing a haze. There is a very close correlation between amount and activity of the vegetation and the concentration of condensation nuclei causing a heat haze or a pollution blanket. According to Went there seems to be remarkably close correlation between the occurrence of asthmatic attacks and the amount of such organic volatiles which are detected with a gas chromatograph in the air.* Whether this turns out to be a casual relationship or not, the enormous amounts of organic matter which are present in the atmosphere and which may play roles in meteorology and pathology should not be overlooked.

PLANTS AS AIR POLLUTANT REMOVERS

Studies into the toxic effects of ozone to plants and in turn the removal of ozone from the air by plants provide an excellent example for the role plants can play minimizing pollution problems. Ozone is one of the principal air pollutants causing damage to plants in the United States (Rich et al., 1970). The uptake of ozone by leaves is essentially an exchange of a gas between leaves and the atmosphere. As such, this uptake must be regulated by the same factors that control the exchange of other gases, such as water vapor. The loss of water vapor from leaves by transpiration is commonly analyzed in terms of resistance. Thus, Rich et al (1970) measured ozone uptake by leaves, converted these data to resistance to ozone uptake, and compared this resistance to water vapor loss. Since the uptake of ozone and transpiration of water are regulated by the same factors, and because realistic simulators of transpiration from natural canopies have been developed (Waggoner et al., 1969), it is now possible to calculate the cleansing of the atmosphere. Laboratory studies and subsequent computer analyses were used to compute how effectively forest canopies could cleanse ozone from a polluted air mass. It was assumed that a mass of polluted air contained 150 parts of ozone per thousand million parts of air (ppb) passes over a forest of trees 4 to 5 m tall. This concentration is usually found on a fairly polluted day in the air of Connecticut. The computer analyses showed that if such an air mass stood over the forest for one hour, the air filtering

* Note by the Editor: As indicated in Part II, Chapter 7, section 2 of this book "Progress in Human Biometeorology", this statement could not be confirmed by other biometeorological studies.

down to the forest floor would have only 60-90 ppb of ozone remaining. The rest would have been taken up by the canopy of leaves. If the polluted air mass stood over the forest for eight hours, the air filtering down to the forest floor would have only 30 ppb of ozone left. Taller trees would remove more pollution than would shorter trees. The larger the stomatal pores and the more numerous the stomata per square inch of leaf surface, the more effective are the leaves in removing ozone from the air.

Damage to tobacco plants, known as "weather fleck", has been recognized in Canada as being due to ozone (Mukammal, 1965). Over the northern shores of Lake Erie, ozone is believed to be mainly generated by photochemical processes from oxides of nitrogen and hydrocarbons emitted in industrial areas to the south. Severe flecking generally occurred with a particular combination of amounts of ozone dose and evapotranspiration. Severity of injury appeared to depend not so much on the absolute magnitude of atmospheric ozone concentration as on the physiological and micrometeorological factors which determine the amount of downward flux of ozone in the lower areas for absorption by the plant. The combination of ozone concentration with the corresponding coefficient of evaporation derived from actual evapotranspiration seemed to correlate fairly well with the degree of tobacco damage, in contrast to absolute ozone concentration and severity of flecking (Mukammal, 1965).

Ozone type injury to tobacco and pinto bean plants increased with increasing humidity (Otto et al., 1969). The size of stomatal apertures increases with increasing humidity. Therefore, humidity may influence plant response to all pollutants and may account in part for the greater sensitivity of plants to ozone type injury in the eastern United States compared with the same species of plants grown in the Southwest.

Ozone, or rather an oxidant defined as the fraction of the atmosphere which will oxidize iodide to iodine and may include peroxy compounds and NO_2 as well as ozone, was proposes as a cause of white pine emergence tipburn (Berry et al., 1963). Susceptible white pines appeared more sensitive to injury by ozone than alfalfa, the most sensitive plant of many field crops and ornamentals tested by Hill et al. (1961).

Where surface water may be contaminated with pesticides and/or other pollutants, percolation of the water through the soil, after prevention of run-off by plants, tends to have a purifying effect (Foy, 1970). Research is in progress to use specific plants to remove and/or detoxify chemicals from water and soil (Foy et al., 1969). These efforts may be divided into three categories: (1) Use of alternative methods of control; (2) minimizing the stages by increased effectiveness and selectivity; and (3) minimizing the stages by removal, inactivation, or alteration of persistance to the chemical once it has been applied.

AIR POLLUTANTS, METEOROLOGY AND PLANT INJURY

Air pollution meteorology and the rather passive role of plants in pollution injury are not considered here. However, surveys have shown that the economic losses, both direct and indirect, caused by air pollution injury to agriculture are extensive and are increasing in many parts of the world. A recent report on plant injury and reduction of yield by non-radioactive air pollutants prepared by a working group of the WMO Commission for Agricultural Meteorology (Mukammal, 1968) reviewed the problem and the present state of knowledge in order to assist meteorologists and others concerned in seeking means of preventing or of controlling the extent of such injury. Several good reviews of air pollution damage to trees have been presented (Keller, 1968; Scurfield, 1960; Hepting, 1964; Hansbrough, 1967). A pictorial atlas (Nat. Air Poll. Admin., 1970) of air pollution injury to vegetation is also available.

PLANTS AS AIR CONDITIONERS

The function of plants as nature's most effective air conditioners and air cleansers is receiving increased attention in our time when there is considerable concern about air pollution and the supply of clean air. The many functions of plants on earth include their role as: (1) Air conditioners by modifying temperature, air flow, and moisture content of the atmosphere; (2) Air purifiers by absorbing harmful excessive carbon dioxide and by giving off oxygen vital to men's survival; and (3) Air cleanser by removing from the air impurities such as air-borne dirt and sand, fly ash, dust, pollen, smoke, odors and fumes (Robinette, 1968). "Plants eat dust" in six basic ways: (1) dilution, (2) precipitation or filtration, (3) narcosis, (4) oxidation, (5) air washing, and (6) deoderization or masking. Dilution refers to the mixing of good or clean air with polluted air. Plant materials and moisture on the leaves trap the dust particles and hold them until they are washed away by the rain. Narcosis is used here as a temporary state of depression, such as wooded areas or forest providing a calm settling chamber for air pollutants. Oxidation is the introduction of oxygen into polluted air. Green vegetation actually removes pollutants and restores oxygen to the atmosphere. The moisture held in suspension on and around a plant serves as an air washer for the polluted, wind-borne air passing around all over a plant and its leaves. Many plants accomplish the replacing of obnoxious odor with a stronger more powerful, bearable or pleasing smell. For example, Korean spice viburnum, honeysuckle, jasmine or mock-orange may be effectively planted between the source and the recipient of a disagreeable odor. Robinette (1968) strongly recommended that fullest advantage of the capabilities of plants as natural air conditioners of men's exterior environment as compared with commercial air conditioners for his interior environment be made.

Although plants are the only supplier of oxygen on earth, the spatial distribution of green vegetation in relation to the concentration of the population on a global scale is not too critical because of the immense reserves of oxygen in the earth's atmosphere and the continuous circulation of air up from the equator, down from the poles and over and across the oceans (Rich, 1968).

It has been suggested that besides green plants bacteria also act as air purifiers. Certain bacteria in the soil help to remove from the air carbon monoxide which is a poisonous gas resulting from men's activities. A large portion of the earth's oxygen supply stems from the phytoplankton in the oceans which occupy three-fourths of the earth's surface. So plants elsewhere in the world still furnish our oxygen when the local vegetation appears frozen and lifeless (Rich, 1968). It appears then that the oxygen replenishment and the composition of air depends on three sources: green vegetation, phytoplankton and bacteria that trap deadly carbon monoxide.

Few statistics are available on the amount of oxygen supplied by plants. Baumgartner (1970) evaluated quantitatively the oxygen supply by forests on earth as compared with the reserves available in the earth's atmosphere of 8,000 meter height. He found that the earth's oxygen reserve is about five million times the daily oxygen production by all forests on earth. In other words, it would take forests some two hundred thousand years to produce the present total oxygen available in the atmosphere. It seems that the oxygen emission by forests is comparatively small in comparison with the immense reserves in the whole atmosphere. Therefore, fluctuations in the oxygen concentrations are only 0.01-0.001 volume percent in the environment of plants and man. The daily vertical oxygen exchange within 1,000 to 2,000 meter above the earth's surface explains why there is no lag of oxygen at night, outside the vegetation period or in areas bare of green vegetation. However, temporary increases in concentration of poisonous gases in densely populated areas is of much serious consequences to man than these small oxygen fluctuations in the absence of green vegetation. It should be kept in mind that the above calculations refer to the amount of pure oxygen. In actual fact, a person breathes daily some 7,280 litres air but reduces the oxygen concentration only from 21-16 volume percent. Therefore, he uses 360 litres or 500 gram pure oxygen. This amount is equivalent to the oxygen emission from some 20 spruce trees in the Ebersberger Forest. The same oxygen amount would be produced by only five of such single standing trees. Every person, therefore, needs the total oxygen in about 1.6 cubic meter air. Since he can use only 73 gram oxygen or the difference between 21 and 16 volume percent, every person needs at least 7-8 cubic meter air for breathing. The inhabitants (1.33 million) of a big city like Munich require daily some 10 million cubic meter air to meet their oxygen requirements. A forest of the same size as the urban area of some 310 square kilometers produces on one day the amount of oxygen contained in 19.35 million cubic meter air. This is about twice the volume of air needed by inhabitants of this area. In other words, the oxygen requirements of a city population are met by high trees occupying about half the city area.

While the above considerations refer to the supply of oxygen by vegetation, Allan (1968) reviewed the literature pertaining to CO_2 fixation in relation to the atmospheric reservoir. Referring to several sources, he cited figures computed by Lieth indicating that the world vegetation could fix the entire CO_2 of the atmosphere in 23-26 years. Thus, plants keep the composition of air in balance by supplying oxygen and removing carbon dioxide. Although short-term fluctuations of CO_2 content may occur during the most rapid periods of growth in the temperature zones, the composition of air on a world-wide basis is kept in balance by plants by supplying oxygen and removing CO_2 in relation to the requirements for air by man and animals.

PLANTS AS AIR POLLUTION DETECTIVES

The important role of plants as research tools, bio-indicators of pollution effects, should not be minimized (Foy, 1970). Compared with animals, plants show a greater sensitivity to smog which makes them excellent early indicators of air pollution (Went, 1966). In addition, the specificity of different plants for various toxic air pollutants makes them usable to monitor air pollution in general and to identify the toxic components even weeks after a pollution episode. This is possible because for each toxic pollutant different damage symptoms appear. The most generally useful indicator plants are leafy vegetables, annual blue grass, and petunias. Sensitive plants are particularly useful as research tools, for example, in studying the influence of relative humidity and other environmental conditions on the severity of air pollution (Otto et al., 1969). The degree of plant injury sustained correlated well with porometer measurements. This indicates that the size of stomatal apertures increases with increasing humidity. Humidity may, therefore, influence plant response to all pollutants.

3. PHYSICAL EFFECTS OF PLANTS ON THEIR ENVIRONMENT

WIND PROTECTION

Problems and technology related to the physical effects of plants on reducing surface wind speed have received considerable attention and a vast amount of literature has been published. Fortunately, several extensive bibliographies and recent reviews of studies related to the effects of shelterbelts and wind breaks on micro-climate, soil climate, soil erosion and plants and animals are available (Chepil, 1965). A working group on wind breaks and shelter belts set up by the Commission for Agricultural Meteorology (CAgM) of the World Meteorological Organization (WMO), prepared a bibliography of literature on the effects of shelter belts and wind breaks on microclimate, soil climate, soil erosion and plants and animals (Van Eimern, 1964). Other reviews pertained to problems of shelter planning which largely depend upon the performance of shelter, the degree of wind reduction of the relevant area and the effectiveness of the types and layouts of belts (Van Eimern, 1968). Nägeli (1954, 1965) studied the effect of a large forest and of the density (permeability), width and shape of scattered shelterbelts on wind reduction. Shelterbelts of different widths but about the same density have similar wind reduction effects, when the distance from the belts is not measured from the leeward but from the windward edges of the belt. Recently, King (1970) investigated positive effects (reduction of soil erosion, improvement of the micro-climate) and negative effects (shadowing) of shelterbelts. Since the negative effects were significant only over a small area close to the shelterbelts but the positive effects over a large area, wind protection of sugar beets resulted in a 3% increase in sugar yields even in a wet and cool year.

SOIL AND WATER CONSERVATION

Whereas crops often need protection against weather hazards, wild plants themselves actively contribute to the conservation of land and water resources (Panel on the Plant Sciences, 1966). Suitable plant covers in watersheds can protect against soil erosion, floods and silted river channels. These reciprocal influences of plants in shaping the environment offer some of the greatest challenges for today's researchers.

During the past decade, considerable progress has been made in the control of soil and snow movement by wind. Following the serious wind erosion damage on the Great Plains of North America in the 1930's, soil surveys and research into soil erosion were accelerated and subsequently numerous shelterbelt projects were established. In Canada, extensive studies by the Soil Research Laboratory at Swift Current, Saskatchewan, included the influence of field shelterbelts on wind velocity, evaporation, soil moisture, and crop yields (Staple et al., 1955). Snow accumulation near shelterbelts varied from almost nil in areas of low snowfall to drifts of 70 feet wide in areas of ample snowfall and strong winds. The net increase in wheat yields in one group of sheltered fields for a 5-year period, taking into account the area occupied by the shelter trees, was 0.7 bu/acre. But sheltered fields showed little net gain from snow when the soil was saturated by autumn or spring rains, when the snowfall was well below average, or when the winds were light and the snow held well on the fields.

Grass sod cover not only helps to stabilize the soil against erosion, but it tends to eliminate or minimize surface run-off, thus conserving moisture and making it available for better crop production. This principle is being used to advantage in current practices of "no-tillage" or "minimum tillage" in crop production (Shear, 1968).

After nearly 30 years of research in the United States to determine primary factors that cause soil erosion, equations have been developed and programmed for computer solution (Skidmore et al., 1970). These programs allow the user to evaluate factors of soil erosion and control practices and to select those that are most economical for his particular field and climate condition.

AIR FILTERING EFFECTS OF FORESTS

Keller (1968) reviewed recent literature pertaining to air pollutants and forests. In general, the filter effects of forests are most effective for dust, aerosols and radioactive pollutants. On the other hand, poisonous gases such as SO_2 are not effectively reduced by forests and can in fact cause a significant reduction in assimilation, damage and eventually death of trees. A concentration of 1 ppm SO_2

reduces assimilation; higher concentrations, even before damage becomes visible, paralyse the normal stomata movement so that SO_2 can be absorbed through the stomata openings in darkness. The density or permeability of forest edges determines the effectiveness for air filtration and the location of dust deposits over the forest stands. About 50% permeability on the windward edge results in greatest dust deposits within the forest (Turner, 1968). A very dense edge, however, produces a stream flow over the top of the forest and dust deposits on top of the canopy, estimated up to 30 tons per hectar for spruce and 60 tons per hectar for beech trees.

Foy (1970) emphasized that plants can be used to minimize pollution problems although very little research has been done on this subject. Woody vegetation has a significant influence in filtering out air pollutants and, especially if conifers and hardwoods are mixed, this could be highly effective in all seasons.

REDUCTION OF TRAFFIC NOISE BY TREES AND SHRUBS

The beneficial role of shrubs and trees in reducing the noise of cities, factories and airports has been recognized, but hardly any research results are available. Plant parts break up sound waves, change their direction and reduce their intensity (Amer. Ass. Nurserymen, 1970). Various types of buffer planting such as hedges, trees, shrubs, vines, and combinations of structural barriers and buffer plantings have been used in landscaping. The noise level above wide expanses of concrete can be reduced as much as three fold if the concrete surface is interspersed with plantings of trees, shrubs and grass. Planting not only decreases noise it also screens moving vehicles from sight, giving a sense of privacy which makes traffic sounds less objectionable. Heavy street traffic registers a noise level of 70 decibles (db). Proper landscaping can reduce the noise about 60%. Through judicious use of plant material, the sound of a heavily highway can be reduced to the level of a sub-urban street in the quiet of evening.

Kämpfer (1967) presented a bibliography on landscape planning in relation to noise reduction. Dense, multi-layer young forest reduced the noise level by approximately 0.16 db per meter; such a forest of 200 to 250 meter extent would reduce traffic noise by 35 to 45 db which is equivalent to the same effect by open fields of 1.8 to 2 kilometer extent (Hess et al., 1962).*

* MEDICAL EFFECTS OF FOREST CLIMATE (Note by S.W. Tromp)

A FOREST CLIMATE may have important clinical effects (see Part II and S.W. Tromp: Medical Biometeorology, Elsevier Publ. Co, Amsterdam 1963: 679-680), because of its effect on the local microclimate surrounding man. It increases the moisture of the air, reduces the number of small ions, increases the number of large ions, reduces the electric conductivity of the air, reduces air movement and local air turbulence (and cooling of the body), reduces direct or reflected solar radiation reaching the body, may prevent formation of ground mist, reduces the natural electric potential gradient of the atmosphere considerably, etc. In other words a forest climate tempers meteorological stimuli on man, which is often favourable but may be detrimental for certain mental diseases.

4. AESTHETIC-HYGIENIC ASPECTS OF VEGETATION

A number of Russian references dealing with aesthetic-hygienic aspects of planted vegetation in so-called "hygienic protection zones" have been brought to our attention. Particularly Rogalina (1960) described the beneficial role of trees and ornamental plants in the vicinity of factories. Recommended standards for the area of sanitary protection zones are given in a series of articles appearing in Gigiene i Sanitariya. Chan-Tzu-Chez (1959) analysed air samples taken in the vicinity of gas filling stations for their presence of carbon monoxide and benzine vapors. He recommended that the distance within the sanitary protection zone from the filling station to the nearest residential area should be 100 meter for stations giving out over 3,000 litres per hour, 75 meters for those distributing from 1,500-3,000 litres per hour and 50 meter with the amount less than 1,500 litres per hour. Similarly, Yanisheva (1969) gave the area of protection zones around electric power stations operating on peat. An electric power station using 90-100 t of fuel per hour with 90% ash recovery should have a sanitary protection zone of 300 m and in the event of 75% ash recovery of 1,000 m. The requirements set up by State standards are sufficient for electric power stations consuming 50 t/hr of low ash fuel and 90% and 75% ash recovery but not for stations using higher fuel rates.

In addition to the recommended standard threshold levels for noise and air pollution as developed in the U.S.S.R., Horn (1963) suggested different permissable maximum values for industrial, residential and recreational areas.

Interesting figures on minimum requirements for the extent of green belts in built up areas were also given by Blancher (1963). He recommended $1m^2$ of planted children's playgrounds per inhabitant, $4m^2$ of sports terrain and $4.5m^2$ of parks. This would require for a city like Paris with 2.8 million inhabitants an area for children's playgrounds of 280 ha as compared with the existing 2 ha; an area of 1,120 ha for sports terrains as compared with the existing 127; and an area of 1,260 ha of parks as compared with the existing 300 ha. Instead of increasing the green belts, as necessary, the present areas planted to vegetation are being transformed into built-up areas. In conclusion, Blancher (1963) stressed that the defence of urban green spaces is a real problem of public health. Their part in atmospheric pollution has been proved as far as certain poisonous gases microbial germs and dusts are concerned. They appear to be very important in the prevention of various psychical troubles, acute and chronic pneumonopathies and also rachitism. He further stressed that the need for green spaces is far from being satisfied in certain large urban districts. Raymond (1963) directed attention to the necessity of maintaining our green spaces in spite of the progress made in the building of large blocks of flats that are much more airy and sunny than those built previously. He refered to the principal causes of atmospheric pollution, chemical and microbial, and insisted on the part that areas planted with trees and various plants that are generally "green spaces" can play both by the screen that they make for protection against impurities, and by the factors of purification they introduce.

5. OUTLOOK

Scientists have given much attention to the various responses of plants to changes in their physical environment. But comparatively little is known about the not so obvious direct effects of plants on their environment budget, "it is the wild plants that influence the conservation of land and water resources" (Panel on the Plant Sciences, 1966). The selection and proper maintenance of suitable plant covers offers practical means for the conservation of soils and water. Although such physical effects of plants on their environment have been intensively studied, much research remains to be done about physical and chemical air cleansing, and aesthetic-hygienic benefits by plants to man in this age of humancentered environment. As recently advocated by the former President of the International Society of Biometeorology (Lee, 1970), biometeorological aspects of new environmental changes should receive first attention in interdisciplinary studies conducted by individuals, study groups, and international societies.

REFERENCES: See pp. 129–130.

CHAPTER 10

METEOROLOGICAL EFFECTS ON ANIMALS IMPORTANT FOR MAN

SECTION 1: Cattle
- (a) Milk Production (H.D. JOHNSON)
- (b) Meat Quality, with Special Reference to the Effect of Heat on Beef Quality (D.F. DOWLING)

SECTION 2: Sheep
- (a) General Effect on Sheep (J.M. DONEY and J. SLEE)
- (b) Effect on Wool Production of Sheep (J.C.D. HUTCHINSON)

SECTION 3: Effects on Pigs (D.L. INGRAM)

SECTION 4: Effects on Poultry, with Special Reference to Disease, Egg Production and Quality (W.O. WILSON)

SECTION 5: Fertility and Reproduction (D. SMIDT)

SECTION 6: Animal Disease (W. FERGUSON and D. BRANAGAN)

SECTION 7: Arthropods (Including Insects, Ticks and Mites) as Vectors of Human and Animal Diseases (W.O. HAUFE)

SECTION 8: Laboratory Animals (W.H. WEIHE)

Meteorological Effects on Milk Production

by

H.D. Johnson

1. INTRODUCTION

Major biometeorological factors which may affect milk quantity and quality are temperature, humidity, radiation, wind velocity, barometric pressure, and rainfall. An adverse biometeorological factor can influence milk production by alteration of the animal's thermal balance, water balance, energy balance, and behavioral activity. These functions are interrelated and involve the neuroendocrine system (Johnson, 1965). Indirectly, these biometeorological factors express their effect on the animal's milk production by altering the quantity and quality of feed and water intake and also may determine whether disease or parasites are present in the environment. In the tropics, meteorological factors of high temperature relative humidity, and indirect factors of inadequate nutrition and parasites is reported to severely limit milk production. Environmental pollution of air, water, and soil by animals themselves or by industry in more urban areas provide chemical modification of the environment which may influence production; however, major emphasis in this discussion will be on the major direct meteorological factors.

Some dairy breeds or species can acclimate or acclimatize more readily to varied biometeorological or climatic environments and thus result in less impairment of the lactation functions. Highly productive *Bos taurus* cattle are commonly crossbred with more heat tolerant and disease resistant *Bos indicus* cattle to increase production and still retain heat and disease tolerance under tropical management conditions. In Tanganyika, Mahadevan and Hutchison (1964) found that helf-breeds were most productive, but in Jamaica the 3/4 Jersey animals were most productive (Lecky, 1950). Although *Bos taurus* have a greater capacity for milk production than *Bos indicus* there are wide breed differences in ability to produce under tropical or temperate conditions in both species or breeds. For example, Mahadevan and Galukande (1962) reported that East African Zebu cows had only about half the potential for milk production of Indian Sahiwal cows emphasizing the genetic variation that exists within the group of Zebu-type cattle.

Work at the Missouri Climatic Laboratory (Worstell et al., 1953; Johnson, 1965) indicate breed differences with *Bos taurus*; however, Johnson et al. (1962) has shown extremely wide individual cow differences in ability to milk in hot climate controlled chambers and (Payne et al., 1957) in natural environments.

2.INFLUENCE OF TEMPERATURE

More recent reviews and reports on meteorological effects on milk yield are those of Bianca (1965) and Johnson (1965a, 1965b). Generally the yield and composition of milk are much the same within the temperature range of 10°C to 20°C. From 20°C to 27°C, the yield decreases slowly and the fat percentage is reduced, but beyond 27°C the decline in yield is much more marked whereas the fat percentage increases and the content of solids non-fat is usually decreased.

Figure 1 describes the milk production of temperate-evolved Holstein, Brown Swiss, Jersey, and the tropically-evolved Brahman cows. Milk production was plotted versus dry bulb temperature of approximately 50% relative humidity and was maximal between 4 and 21°C. Milk production generally declined above and below this comfort zone. More specifically at the high environmental temperatures, the yield declined for Holstein animals at approximately 21°C, Brown Swiss and Jersey at about 24-27°C, and the Brahman at approximately 32°C. The most pronounced decline in lactation was shown by the Holstein animals. Table 1 illustrates the relative effects of temperature on milk production, feed intake and the conversion of feed to milk by these four breeds.

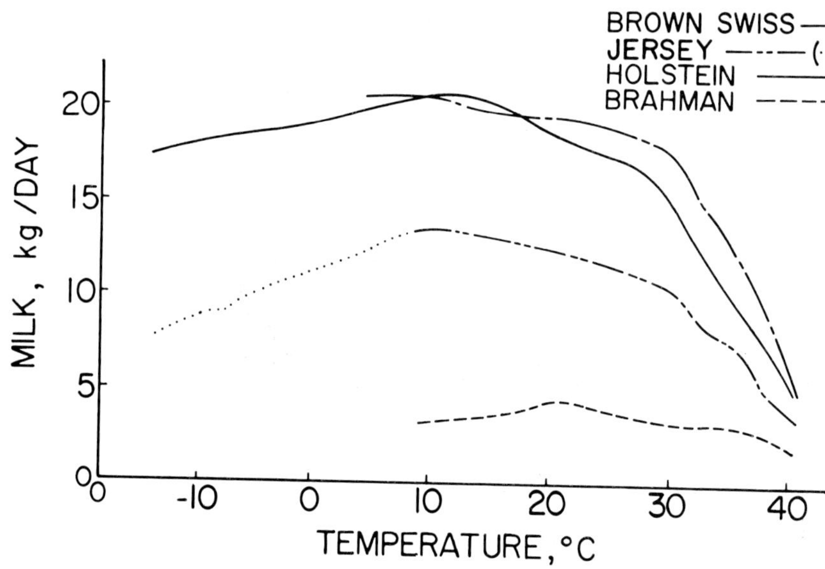

Fig. 1. Comparison of actual average values of temperature effects on milk yield of cattle in controlled temperature conditions of the laboratory. Relative humidities ranged from 40 to 60% in this range of temperature. Based on data from Ragsdale et al. (1950, 1951).
(*Low temperature values – dotted line – for Jersey cattle are based on different animals than the group used for the values above 10°C).

Table 1. Effects of -13°C as compared to 10°C on milk production, TDN intake, and the ratio of conversion.

	Milk, kg/day		TDN kg/day		Ratio feed/milk	
	10°C	-13°C	10°C	-13°C	10°C	-13°C
Holstein	15	14	10.4	11.2	.70	.80
Brown Swiss	—	—	—	—	—	—
Jersey	7.6	3.5	5.9	7.4	.78	2.1
Brahman	—	—	4.5	6.2	—	—

Effects of 38°C as compared to 10°C on milk production, TDN intake and ratio of conversion:

	10°C	38°C	10°C	38°C	10°C	38°C
Holstein	18.6	4.9	12.1	2.6	.65	.53
Brown Swiss	20.4	9.5	12.2	2.5	.60	.26
Jersey	13.5	5.1	8.8	2.8	.65	.55
Brahman	3.6	3.2	5.2	3.3	1.44	1.03

$$\frac{\text{Feed kg/day}}{\text{Milk kg/day}} = \text{kg of feed (TDN) required to produce 1 kg milk.}$$

Note: The data for Holstein and Jersey cows are from Ragsdale et al. (1950). The data given for Brahman cows at – 13°C are from Ragsdale et al. (1950) and at 38°C from Ragsdale et al. (1951), and the Brown Swiss data from Ragsdale et al. (1951).

The following selected references indicate the extent of cold and hot temperature investigations on cattle. Earlier work by Davis (1914) reported a lowering of milk yield of cows held in both open and closed barns during cold weather and also observed that cows in the open sheds consumed more roughage.

Popoff (1972) found that cows produced less milk at a lower temperature, –10 to 4°C, than in a higher temperature of 5 to 12°C. Kelly and Rupel (1937) observed that changing cows from a warm barn to an open lot (average temperature of –8°C) resulted in a daily decrease of 0.38 kg/cow. However, extra feeding minimized this effect. Dice (1935, 1940, 1942) reported similar results on cows and heifers. Witzel and Heizer (1946), in comparing FCM milk of cows kept in cold barns with cows kept in warm barns, showed the cold animals to produce within 2.7% of the warm group. Sementovskaya and Garkavi (1950) exposed 8 cows to outside winter cold and without supplementary feeding observed a 26% decline in milk yield and a 28.5% increase in butterfat accompanied by a loss in body weight of approximately 0.9 kg/day. All of these publications suggest that cattle may lactate successfully at cold temperature but with lowered production, and greater feed intake.

Regarding high temperatures, earliest reports were those of Regan and Richardson (1938) who observed in Holstein, Jersey and Guernsey cows a 40% drop in milk production when the temperature was increased from 4 to 35°C. Early studies in the Missouri Climatic Laboratory (Ragsdale et al., 1948) indicated a decline in milk production at 24-27°C for Holstein, 29°C for Jersey cows, and 29°C for Brown Swiss cattle (Ragsdale et al., 1951). Later studies described the response of milk yield to high temperature and the influence of heat tolerance acclimation and the level of milk production on the heat effect (Johnson et al., 1960; Johnson et al., 1962; Johnson et al., 1967). The breathing of air cooler than the environment temperature has been found to be significant in preventing a decline in milk production (Hahn et al., 1965).

Thus, most information points to the depressing effect of unfavorable summer environment for milk production in most temperature zones. An exception is a recent report by Kali et al. (1968). However, the seasonal comparison of these data are confounded with superior feeding and high levels of management, such as spraying animals with water during the summer. Valid interpretation of the seasonal effects would be possible if the immediate micro climate of the animals during summer and winter were reported. Also McIntyre (1967) reported no seasonal effect on milk production; however, winter feed intake was limited which lessens the effectiveness of a direct seasonal comparison. Since rectal temperatures were not elevated and milk production was rather low, temperature was apparently not a major limiting factor for these animals.

Some effects of temperature on composition are those of Sargent et al. (1967) in North Carolina in which lactation initiated during the summer months of July, August, and September was lowered; whereas solids nonfat (%PLM, protein-lactose-mineral) were higher than cows freshening in November, December, and March. Fat percentages were similar to the PLM trends. Gacula et al. (1968) reported similar trends in Massachusetts. Most workers have reported that the solids-not-fat (SNF) content of the milk decreases in the heat-stressed cow. Regan and Richardson (1938) gave values of 8.26, 8.06, 7.88, and 7.58 at environmental temperatures of 4, 15, 27, and 37°C. Parallel to this there was a decline in casein percentage. Decreases in the SNF percentage of the milk of cows of various breeds have been found to occur at temperatures higher than about 30°C (Cobble et al., 1951), as well as in cows that had elevated body temperatures owing to various conditions of disease (King, 1955). Protein percentages in the milk of cows of various breeds in the humid tropics have been reported by Bateman and de Alba (1961). The values ranged from 3.29% in Sindhi x Brown Swiss to 3.59% in Criollo cattle. Most work clearly demonstrates a decrease in SNF at higher environmental temperatures and some tendency toward elevation at colder temperatures (Cobble et al., 1951; Davis et al., 1914; Heinemann, 1947; Houston et al, 1932; Moody, 1967; Richardson, 1960). Richardson (1960) and Richardson and Johnson (1961) presented evidence that absolute butterfat production paralleled the declines observed in milk yield at the high temperature and humidity levels. A depression in absolute milk fat production (kg/day) occurred above 20°C. The percentage of butterfat tended to increase at the higher conditions. Richardson and Johnson (1961) and others showed that any combination of temperature and humidity above 18°C may cause definite changes in the composition of milk fat. Fatty acids of lower melting points such as short chain ($C_6 - C_{12}$) and oleic acid decreased at the higher environmental temperature and three fatty acids (palmitic and stearic and myristic) of higher melting points increased at the higher environmental temperatures.

Environmental temperature studies on composition of milk protein have mainly been indirect estimated by use of total nitrogen values. Most workers (Cobble et al., 1951; Richardson, 1960) have found the total nitrogen of milk to decrease at the higher environmental temperatures and increase in cold (Merilan et al., 1959).

Richardson (1960) also observed the NPN of milk to increase under thermal stress which suggested the action of gluconeogenesis. Temperature affects the milk salt balance, citric acid, phosphorus, calcium, magnesium, and potassium. These values were lower at a high temperature 27°C and 32°C, than at a low (18°C) environmental temperature. The average citric acid content was higher at 18°C than corresponding values at 27°C for early, mid, and late lactation periods. At high environmental temperatures (27°C) the total phosphorus content in milk was markedly depressed. Magnesium content at 27°C and 32°C were lower than those at 18°C throughout most of the lactation period. Total calcium content was lower at 27°C and 32°C than at 18°C during early lactation but became higher during the rest of the lactation period. At high environmental temperatures, 27°C and 32°C the salt balance was markedly lower than the corresponding values of 18°C. These temperature effects are of practical significance to the diary industry in the process of manufacture of evaporated milk. To lower citrates and phosphates in proportion to calcium and magnesium cause heat coagulation in sterilization and pasteurization. Milk leucocyte counts increase markedly under heatstress conditions (Nelson, et al., 1967) and abnormal milk due to thermal stress was reported by Roussel et al. (1969).

DIURNAL TEMPERATURE CYCLES simulation on milk production were shown to compare similarly to responses obtained at constant temperatures equivalent to the mean diurnal cycle temperature (Brody et al., 1955). A temperature cycle of 21-38°C depressed production of Holstein cows by 20% and Jersey cows by 8%. As suggested by Johnson et al. (1960), tolerance to heat is modified by level of production and duration of heat exposure. And when additional factors of nutrition are included some inconsistencies in the temperature effects on yield are reported. Generally, if the hot micro climate is severe enough to reduce daily feed intake, the milk production will decrease (McDowell et al., 1969). Data of Johnson et al. (1961) support this generalization but force-feeding the animal by rumen cannula did fully compensate for the heat induced lowered metabolic activity of body and mammary glands.

3.INFLUENCE OF HUMIDITY

Milk yields are affected by humidities at temperatures above 27°C (Johnson et al., 1963). To quantify these relationships a temperature-humidity discomfort index (Johnson et al., 1962b) was established for milk production. An environmental index that quantitates the contributing effects of radiation, humidity, wind and diurnal cycles, on lactation is of course more meaningful since dry bulb temperature is always accompanied or qualified by the other variables. Such an environmental "equivalence index" should be established. An index of this nature would relate more directly to the heat balance and comfort of the animal than dry bulb temperature. The application of the temperature-humidity index to milk production was an effort in this direction (Cargill et al., 1962; Johnson et al., 1962a, b).

Figure 2 illustrates the effects of various temperature-humidity conditions on animals producing approximately 23, 18, 14 kg/day. Using 18°C as a base condition the decline in milk production at each THI condition was computed from the following equation: (THI = t_{dp} + $.36t_{dp}$ 41.2). These values were based on declines in production from base (18°C) values. A switchback procedure was used for the various temperature humidity conditions thus eliminating the persistency

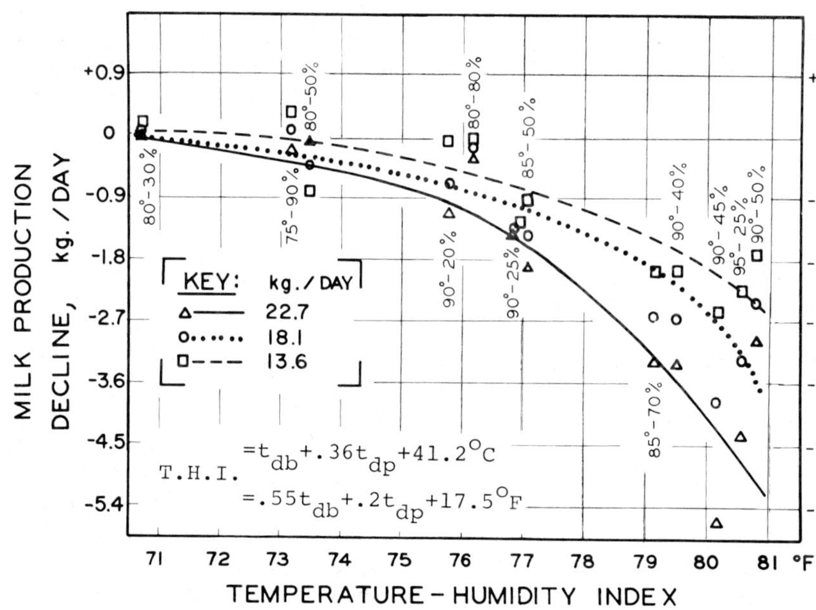

Fig. 2. Declines in milk production from base 18°C values (switchback design) at various THI conditions. Data from Johnson et al. (1961, 1962) and Cargill et al. (1962). The original equation (Weather Bureau, 1959) was in Fahrenheit so all values in this figure are so expressed.

effects on lactation. THI values have not been established for conditions below 18°C for cattle. Below 18°C a THI index may be of limited significance; however an index, relating radiation, wind and humidity to temperature would be a most significant contribution on lactating cattle at a cold temperature. The previous application of Wright's climograph (1946) was a valuable contribution in this aspect of shelter evaluation. More recently Hahn (1969) reported that these data as a valid base for predicting production losses in various regions in the U.S. Expected production losses then provide a rational basis for making decisions among environmental modification alternatives (Hahn et al., 1969; Hahn et al., 1970).

4.INFLUENCE OF WIND

Wind or ventilation increases convective and evaporative exchange of heat between the animal and the air about it. Data from the Missouri Laboratory show that increased wind at temperatures below the comfort zone may become detrimental to milk production. The wind shifts the comfort zone to higher temperatures. The data of Table 2 demonstrate that wind is advantageous at higher temperatures. Air velocities of 5 and 9 mph proved beneficial in recovering milk production at an air temperature of 35°C that was otherwise lost when the velocity was only 0.4 mph (Brody et al., 1954b).

Table 2. Effect of wind on milk production at four temperatures.

	Milk production – % of Normal					
	-8°C*			10°C*		
Wind mph	Holstein	Jersey	Brown Swiss	Holstein	Jersey	Brown Swiss
0.4	76	36	72	100	100	100
5.0	85	39	74	100	100	100
9.0	84	35	75	100	100	100

	27°C*			35°C*		
Wind mph	Holstein	Jersey	Brown Swiss	Holstein	Jersey	Brown Swiss
0.4	85	100	100	63	74	83
5.0	95	100	100	79	94	90
9.0	95	100	100	79	94	90

* Relative humidity range, 60-70%.

5. INFLUENCE OF RADIATION

When the air is hot, radiation will augment the decline in milk production. Exposure to 180 Btu hr^{-1} ft^{-2} did not affect milk production at 7°C, but did cause a decline at 21°C and above (Brody et al., 1954a). The data of Table 3 verify this: at 28°C the decline in milk produced by either Holstein or Jersey was sharpest when radiation was greatest. The difference in the animal's response to shades at different locations points out the importance of radiation in the micro climate and the animal's energy budget. Shade does not effectively change either the air temperature or humidity at a particular location, but it does change the radiation items in the energy balance of an animal (Bond, 1961). This change is significant in areas where shade affects production. Its significance will not depend upon high air temperature alone. In the dry climate of the Imperial Valley of California, parts of the sky are heat sinks with effective temperatures, for example, −2°C lower than the air temperature of 38°C (Kelly et al., 1937). Here the effect of an efficient shade would block out the "hot spot" of the sun and leave the animal exposed to the "cool spots" of the sky. Shades are practical in many regions, but their effect may differ by degree in different regions. Brunt's formula relates incoming long-wave radiation from the air to vapor pressure at the earth's surface. This shows that the effective temperature of the sky is greater when the humidity is higher. In a humid region "cool" areas in the sky may be nonexistent or much warmer than in a dry area with the same air temperature. Thus, Bond (1961) suggests caution in transferring results from one climatic region to another. A recent paper by Hahn and Osburn (1970) discusses this subject.

Table 3. Effect of radiation on milk production at three air temperatures.*

Level of radiation Btu ft $^{-2}$ hr $^{-1}$	Milk production –% Normal 8°C		21°C		27°C	
	Holstein	Jersey	Holstein	Jersey	Holstein	Jersey
Variable	100	100	99	99	84	94
5	100	100	100	100	93	99
40	100	100	100	99	92	94
90	100	100	93	95	77	93
130	100	100	90	97	69	87
180	100	100	88	95	57	79

* From a paper by R.G. Yeck, 1958 (Presented at the Third Annual Farm Electrification Short Course).

6.INFLUENCE OF ALTITUDE

Environmental temperatures vary with altitude in Central and South America as in all tropical countries (Montero, 1966). At the higher altitudes of these countries, with moderate temperatures and improved nutrition, higher producing European breeds of cattle are reasonably well adapted; however, I am not aware of any quantitation of altitudes effects on milk production.

The higher temperatures and associated humidity and radiation, poorer natural forages, and diseases in the lower altitudes or coastal regions dictate the use of hybrid milk cattle or breeds indigenous to the tropics. Associated with the greater adaptability of these animals is much lower production of milk.

SUMMARY

With the trend toward dry lot and confinement housing, it is more imperative that we know the optimal levels, limits and risks of various meteorological factors as they relate to milk production. The influence of these meteorological factors and their indirect consequences contribute to the overall physiological and management efficiency of milk production. To my knowledge there are no quantitative data on the effects of photoperiodism, air pollution, parasites and disease on milk production. There are other meteorological factors not discussed that may prove to be limiting to milk production in future management systems. As we improve the genetic potential of our animals to produce milk, we must have scientific data to prescribe the most optimal meteorological environment for high milk production. In the future the need for greater efficiency in milk production will become increasingly a problem expecially in developed and underdeveloped heavily populated areas of the world.

REFERENCES: see pp. 131–133.

Meteorological Effects on Meat Quality With Special Reference to the Effect of Heat on Beef Quality

by

D.F. Dowling

1. INTRODUCTION

Meat quality, or what the consumer wants, is difficult to define. The demand has varied at different times and according to customs. With beef for example it has been associated with the type of animal which is usually associated with compact animals, excessively fat and relatively pampered. The emphasis on intramuscular fat, as a measure of quality, has led to slaughtering of cattle with excess fat on the one hand and the popularity of small, early maturing beef breeds on the other. Such animals are uneconomical in temperate regions yet have been tried under hot tropical conditions for even greater loss. Cattle breeding in the tropics is giving encouraging results when functional efficiency in the environment used, is the basis to the evaluation of cattle.

Previously, cattle which were found in the tropics, especially those under stress due to lack of animal care and management, were reported of poor conformation and thought not suitable for beef. Usually beef produced under such conditions is tough. Beef may be a rare and expensive dish in the traditional serves. Thus before Charles (1964) developed a practical system to classify beef carcasses on the basis of sex, age, weight and fat cover, conformation featured prominently in beef grading systems. Conformation was thought to indicate beef quality and the proportion of more valuable beef cuts in a carcass. Beef type cattle are thought to be superior to dairy type cattle in these attributes and consequently attracted a higher grade. However, Butterfield (1963) found that there was little variability in the muscle weight distribution between breeds. Further Berg and Butterfield (1966) showed that excess fat had a large influence on conformation. Subcutaneous fat evened the outline – improving the apparent "beef" type of an animal but in actual fact decreased the retail yield of muscle content. Berg (1966, 1969) has emphasised that this type of animal, mistakenly produced for "quality" beef, is uneconomical and results in less beef compared with alternative crossbreeding methods. Hence the aim in cattle breeding is to breed in the carcass and adaptability qualities required to produce edible beef more cheaply (Dowling, 1956, 1970). Beef production in the tropics can only but progrress if the presently available knowledge is applied. The liveweight curves of cattle in the tropics in North Australia in Figure 1 illustrates the continuous growth provided the grass is available. Similarly the graph in Figure 2 reflects the influence of nutrition rather than meteorological factors.

Also it demonstrates the significant genetic variability in growth and adaptability in a tropical environment.

2. INFLUENCE OF TEMPERATURE AND HUMIDITY

Climate can perhaps be regarded as the primary factor determining beef production. It determines the vegetation which is an integrator and expression of the total environment. However, our attention in this paper is focused on the micro-climate and is restricted to one factor, viz. the effect of heat.

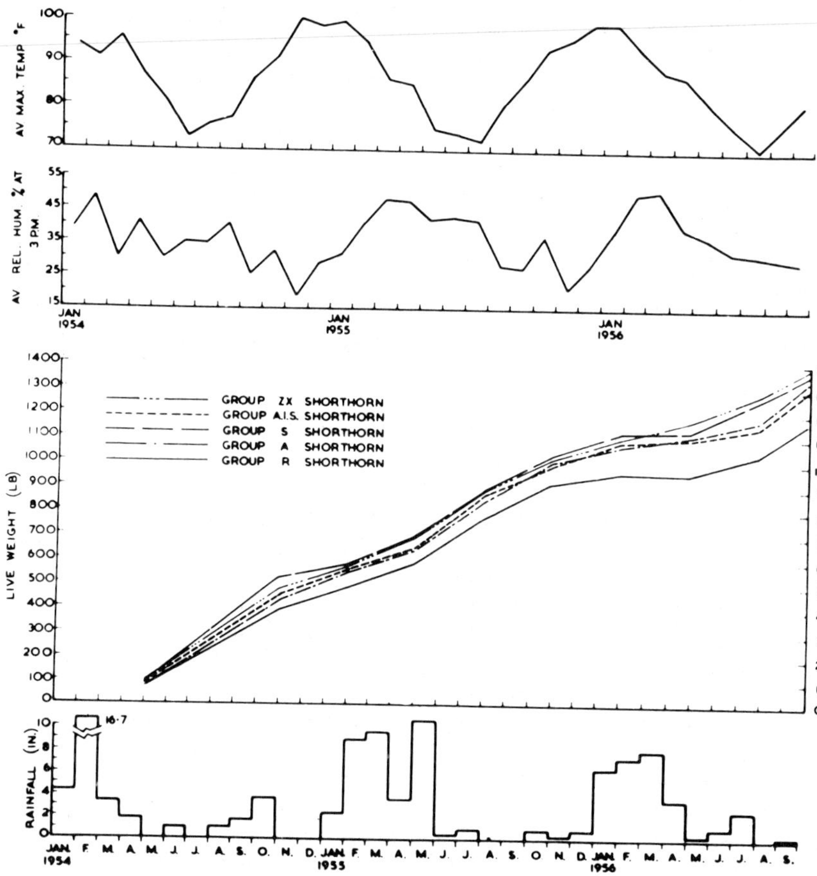

Fig. 1. Growth curves of three strains of cattle in a tropical environment in North Australia. The growth curves follow the rain and availability of forage rather than meteorological factors, viz. temperature, etc.

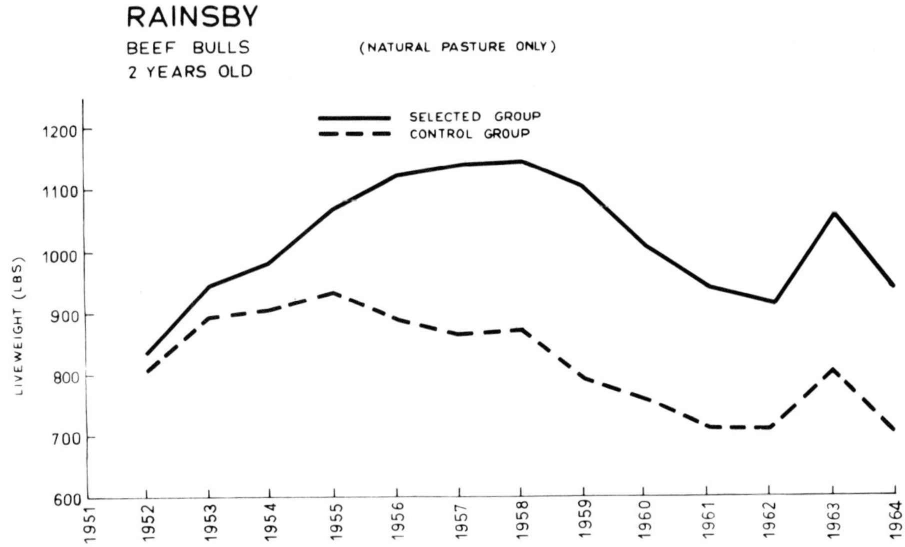

Fig. 2. The growth curve illustrates the effect of selection with a breed of *Bos taurus* cattle within a tropical environment can be significant. An infusion of *Bos indicus* makes for further gains.

The effect of high atmospheric heat especially if accompanied by high humidity or other factors which make up weather heat stress would tend to affect the appetite and growth. However, normally hot atmospheric temperatures per se need not suppress production. Provided ample water and fresh nutritious green forage are available in an environment, i.e. low humidity and air movement which enhances the effective vaporisation of the sweat, cattle need not be unduly stressed (Dowling, 1956a). Conversely, Webster (1969) has demonstrated that cattle can adapt to intensely cold conditions of -11° to -22°C.

The tropics have high temperatures throughout the year. Cattle will grow steadily throughout the year too, according to the freedom from stress and the level of nutrition. Consequently the overall greater yield is possible. Similarly cereal crop yields are low in the tropics compared with yields in temperate areas, but in the tropics more than one crop can be grown per year for an overall higher production. Certainly grass forage yields are also higher but until recently appeared to be less digestible and of lower value.

Cattle have a remarkable capacity to sweat efficiently when adapted to extreme heat conditions (Dowling, 1956b) (Fig. 3). Cattle adapt to extreme environmental conditions or succumb. The type of cattle which have adapted to harsh hot environments may not make the best use of pastures for beef. Though beef qualities like tenderness, beef flavour and aroma, etc. are not associated with growth rate. No advantage of quality accrues for young beef type animals over similar dairy type cattle. Similarly the tick and disease resistant animals in the hot, humid areas on rank, imbalanced forages may not be suitable to make full use of pastures for beef. Consequently, the consumer's chance of obtaining tender meat in tropical conditions must be improved. This can be done by crossbreeding for the desired carcass type of animals for optimum production under improved pastures. Better management practices may well demonstrate that the greatest potential for beef production lies in the hot tropical regions when more efficient animals will be bred for the tropics. Mostly, cattle for beef were supposed to be under more or less continuous stress and that the odds were insurmountable (Dowling, 1970). There was a time too when it was though that "white" man could not adapt to the heat. Certainly the effect of weather factors on the incidences of parasites, diseases, and dis-eases

Fig. 3. Shows experimental cattle in the extremely hot, dry, shadeless regions of tropical Australia. Growth rates of cattle may be recorded under these conditions equivalent to those recorded elsewhere under grazing conditions. Naturally occurring heat stress does not constitute an acute problem of survival for cattle provided heat dissipation by the vaporisation of sweat is enhanced (e.g. water and air movement). The *Bos indicus* infused group averaged 17.3 Kg more carcass at 2 years but the selected *Bos taurus* group averaged 81.8 Kg more carcass than steers compared in the temperate regions (due to periods of pasture dormancy, etc.).

which may suppress beef production, though more difficult to identify precisely, is easy to see in cattle uncontrolled in the tropics.

Grazing industries based on tropical grasslands were of low productivity in terms of both quality and quantity. Whyte (1962) considered that there was only limited scope for improved pastures but he did stress the need for further research. In the higher rainfall areas, this research was forthcoming. Griffith-Davies and Eyles (1965) and Hutton (1970) have shown that cattle numbers can be trebled for five-fold increase in production with soil fertilisers and balanced pastures. There are millions of acres in the tropics alone where the soil fertility is such that the normal agricultural procedures for utilisation are not possible or economically feasible. In the process of improving this land by methods accrued by research on soil and legume pastures, animals could well be grazed beneficially. Certainly this has been amply verified in tropical Australian grazing conditions where beef can be produced as choice, tender and flavourful as anywhere in the recognised beef areas (which are subject to periods of pasture dormancy and high costs of fertilisers and mechanisation).

3. INTENSIFICATION OF MEAT PRODUCTION AND QUALITY

Engineering science can overcome climatic factors at both extremes of cold and heat. There is no question that high intensified systems of animal production can work in hot areas. Of course intensification entails sufficient capital and equipment, the application of knowledge to obtain economical and beneficial results which can only accrue with the trained personnel. Confinement of animals in most instances has improved performance and "quality" considerably (dairy, calves, pigs, poultry, broilers). At the same time confinement and closer animal contact increases the possibility of disease and stress problems. Specialisations and trends towards overcoming animal problems of full confinement are obvious. More recently the critical problems are those associated with the handling of animal wastes and management factors of a biometeorological nature.

CONCLUSION

Biometeorological effects on meat quality are not easy to ascertain. Meteorological factors need not directly affect the quality of meat of animals under proper nutrition, sound management, with suitable animals under a positive animal health and production regime. Stress undoubtedly affects the tenderness, flavour and juiciness of meat. Evan fat colour is affected by pre-slaughter stress and post-slaughter conditions. For example, high humidity and slow chilling in still air make the fat more yellow. Advanced technological care of animals is required even under experimental conditions to ensure maximum results. We are concerned with optimum results with the minimised costs in the production of quality in meat. With the control of disease and the *Bos indicus* type of cattle infused with the desired

Fig. 4. Shows the results of infusing pure *Bos indicus* (Sahiwal) in beef Shorthorns selected for growth, fertility and adaptability in a tropical area. Inter se matings of F_1 generation animals produced the "Quasar" strain shown above. The bulls weighed over 680 Kg at 18 months of age.

carcass composition quality, beef production under hot conditions is highly successful (Fig. 4). Similarly, *Bos bubulos,* the water buffalo, demonstrates the choice, flavourful meat that can be produced under still more stressful tropical conditions. Hence it would appear the meteorological effects on meat quality can be overcome anywhere under intensified conditions if costs are not limiting. Thus the world's greatest potential for meat production lies in the vast areas of the humid tropics with year round warm weather, high rainfall and deep porous soils. Further in many more millions of hectares in the drier regions of the tropics, the soil fertility could be improved for pastures for cattle where meteorological effects on meat quality can be biologically controlled at much less expense.

REFERENCES: see pp. 133—134.

General Meteorological Effects on Sheep

by

J.M. Doney and J. Slee

1. INTRODUCTION

The purpose of this review is to describe the influence of climate on the physiology, survival and productivity of sheep. Exposure to extreme heat or cold causes the sheep, as a homeotherm, to respond in such a way as to maintain its body temperature. These responses, which may be modified as a result of acclimatization, adaptation or genetic variation, will influence the impact of climate upon the sheep in terms of its survival and productivity. Survival in the heat requires that the heat load shall be equalled by heat dissipation before body temperature (normally about 39.5°C) rises excessively. Body temperatures of over 42°C can be tolerated by sheep, but probably for short periods only. Under cold stress the requirements are for heat conservation to be maximised and for heat production to balance heat loss. In practice these conditions are generally satisfied without body temperatures falling below about 38°C (Bligh et al., 1965) although much lower temperatures can be tolerated (Slee, 1966). The husbandry requirement is that these equations should be balanced with the minimum effect upon productivity.

2. PHYSIOLOGICAL RESPONSES OF SHEEP TO HEAT AND COLD STRESS

The major physiological responses of the sheep to HEAT STRESS are vasodilatation on the extremities (encouraging sensible heat loss), coupled with both sweating and panting which involve evaporative heat loss. According to Webster and Johnson (1968) the ears and legs of Merino sheep contribute about 23% of the body surface area and dissipate a high proportion of sensible heat up to environmental temperatures of about 36°C. However, when the environmental temperature approaches mean skin temperature (about 37-38°C) sensible heat loss by convection or radiation becomes unimportant and evaporation becomes the main avenue for heat dissipation. For sheep, unlike cattle, sweating is much less important than respiratory evaporation (Brockway et al., 1965). In unshorn sheep sweating may in fact affect thermal balance adversely because of the exothermal reaction produced when water is taken up by wool (King et al., 1940; Bligh, 1967). However, sweating on the scrotum may be effective in cooling this region (Waites et al., 1963). Respiration rates in sheep vary from 14-16/min at subcritical or thermoneutral temperatures (Webster et al., 1968) up to over 300/min at high ambient temperatures (Riek et al., 1950) when respiratory evaporation is almost the only source of heat loss.

The work of Parer (1963a, b) and Eyal (1963) has shown that the fleece impedes heat flow whether in the form of radiation or convection. Thus when the ambient temperature is higher than skin temperature and the sheep is exposed to radiant heat from the sun, the fleece may reduce the heat load so that shorn sheep have higher respiratory rates and higher body temperatures; on the other hand when the animal is in shade, the main effect of the fleece is to reduce heat loss and then shorn sheep may have the advantage.

Physiological and behavioural responses to COLD STRESS are either energy-saving (e.g. peripheral vasoconstriction, lowered respiratory rate, postural changes and shelter seeking), or energy-consuming (shivering coupled with increases in heart rate and heat production). Physical parameters such as body size, tissue conductance and coat type and length interact with the climatic variables (temperature, wind speed, rain and radiative environment) to determine the physiological responses necessary to maintain body temperature. These interactions have been discussed by Blaxter (1964a, b), Joyce and Blaxter (1964) and Joyce et al. (1966).

The EFFECTS of WIND, RAIN and RADIATIVE HEAT LOSSES can be equated in terms of the total cooling power of the environment. Wind and rain are different only in so far as they break down fleece insulation (Blaxter, 1964a; Doney, 1963), which is an important component of the sheep's ability to resist the stress of cold climate. Long fleeces afford most insulation but are proportionately more affected by wind (Joyce et al., 1964; Griffiths et al., 1969). Possibly the dense Merino fleece gives most insulation per unit length (Bennett et al., 1964), whereas longer fleeces may be more resistant to rain (Blaxter et al., 1966; Doney, 1966; Doney et al., 1968). Alexander (1962, 1964) indicated that long birthcoats in newborn Merino lambs afford better insulation and so affect survival. Purser and Karam (1967) and Purser (1967) reported that differences in the mortality of newborn Welsh Mountain lambs during bad weather are associated with birthcoat morphology.

MORTALITY AMONG LAMBS EXPOSED TO COLD STRESS in the field is common (e.g. Alexander, 1964; Obst et al., 1968) but may be alleviated by shelter (Watson et al., 1968; McLaughlin et al., 1970). However, Miller (1968) found that shelter was used more by young lambs than by their grazing dams, so that the frequency of suckling was reduced. Sometimes adult sheep die from the cumulative effects of cold and undernourishment during long severe winters (Blaxter, 1964b) or as a result of acute cold stress (usually involving wind and rain) following shearing (Hutchinson, 1968). Jute coats augment fleece insulation in wet, windy weather (Webster et al., 1967) and plastic coats reduced mortality among shorn Merino ewes exposed experimentally to cold, wet conditions (Panaretto et al., 1968). Poor nutritional status increases the likelihood of mortality, both in lambs (Alexander, 1964) and in shorn adult sheep (Hutchinson, K.J. 1966 and Hutchinson, J.C.D. 1968). However, even sheep with adequate reserves may die as a result of short, acute cold stress (Panaretto, 1968). Such deaths, some of which involve a degree of hypothermia less severe than that known to be tolerable (Slee, 1966), may be attributed to adrenal insufficiency consequent upon excessive ACTH secretion (Panaretto et al., 1969). Differences in plane of nutrition also influence the ability of sheep to resist body cooling in climate chambers (Slee et al , 1967; Sykes et al., 1969a, b and Wiener et al., 1969). Finally, there may be a sex difference (female superiority) in cold resistance under certain conditions (Slee, 1970a).

HEAT STRESS MORTALITY is probably rare in sheep since ambient temperatures of over 44°C (47% RH) can be tolerated for daily periods of 9 hrs, at least in Merino sheep (Alexander et al., 1971). Hot climates may of course increase mortality among sheep indirectly as a result of drought and food shortage. Under conditions of heat and drought, economy of water utilisation is a crucial factor (Macfarlane, 1964)

3. ACCLIMATIZATION AND GENETIC ADAPTATION

The effectiveness with which sheep can tolerate or survive cold may be influenced by ACCLIMATIZATION. For example, the ability of shorn sheep to resist hypothermia during severe cold stress can be increased by previous acclimatization resulting from chronic cold exposure or acute cold stress as reported by Slee and Sykes (1967), Sykes and Slee (1968 and 1969a, b). Acclimatization has also been induced in unshorn sheep by natural outdoors exposure (in Canada) and by equivalent dosages of cold in climate chambers (Webster et al., 1969). Acclimatization usually involves increased resting metabolism and heat loss and improved heat production capability and is, therefore, energy-consuming. Some forms of natural acclimatization to fluctuating temperature or wind (Webster et al., 1969; Griffiths et al., 1969) or habituation to short cold shocks (Slee, unpublished) may however produce an energy-sparing type of response which would be appropriate for sheep exposed to prolonged cold stress with undernutrition.

There is little evidence for physiological acclimatization to heat in sheep although Wodzicka (1960) reported that the body temperature of rams was elevated less by experimental exposures to heat in summer than in winter and Alexander and Williams (1971) found that the distress of sheep in the heat diminished after several days exposure.

GENETIC FACTORS, including breed differences, may influence the sheep's ability to withstand extremes of temperature. Blaxter et al., (1966) tested six breeds and found differences in heat production in response to wind and rain. These differences were attributed to fleece type. On the other hand Slee (1966 and 1970b) and Sykes and Slee (1969a) found considerable variation among shorn sheep in ability to resist experimental body cooling both within breeds and between breeds. Individual differences were highly repeatable. Current work (Slee and Purser, unpublished) suggests that cold resistance may have a heritability of about

35% in the Scottish Blackface breed. Joyce and Blaxter (1964) reported a difference between Down cross sheep and mountain breeds of sheep in tissue insulation, but this was subsequently not confirmed (Webster et al., 1966). Slee (1968) and Griffiths and Doney (1969) found that Scottish Blackface sheep vasoconstricted more readily than Tasmanian Merinos during cooling, thus producing differences in heat conservation. In the field, Obst and Day (1968) found that the mortality rate among Merino lambs significantly exceeded that in Corriedales during periods of wind and rain. Similarly, different genetic lines produced by selection in the Welsh Mountain breed consistently show differences in perinatal mortality during bad weather (Purser et al., 1967).

BREED DIFFERENCES IN HEAT TOLERANCE probably also exist. Symington (1960) compared German Merinos, Blackheaded Persians and indigenous Rhodesian hairy sheep in a tropical environment and found that the Merinos had slightly higher body temperatures despite more sweating and higher respiration rates. Fowler and Dun (1966) and Fowler (1968a) found differences in physiological response and fertility among heat-stressed Merinos previously selected for and against degree of skin fold. Shafie and Sharafeldin (1965) studied heat tolerance in Ossimi, Texel, Caucasian Merino and German Merino breeds. The indigenous Ossimi was superior in terms of rectal temperature and respiration rate and the Caucasian Merino was the best of the foreign breeds.

4. PRODUCTIVITY

Sheep inhabit a wide range of climates from the sub-tropics to cold uplands Moreover they can utilise low quality roughage produced by poor land, which might otherwise be unexploited. Hence the traditional sheep environment is an extensive grazing system with animals fully exposed to the prevailing climate. Productivity is therefore affected by climate and by the physiological responses induced by short term changes in weather.

Sheep indigenous to hard environments are thought to have low genetic potential for production; hence the emphasis on new breed formation utilizing improved genotypes. The maintenance of "hardiness" or genetic adaption is also relevant (Kurz, 1955).

Comparisons have been made within both hot and cold environments between introduced breeds, cross-breds and indigenous sheep (Kijatkin et al., 1959; Symington 1960; Shafie et al., 1965; Taneja, 1966; Seth et al., 1966; Tyagi et al , 1969; Nasyrov, 1969) and the performance of introduced populations in their new environments has been compared with that in their established areas (Alekseeva, 1962; Popovici et al., 1966). The overall performance of up-graded genotypes has generally been found superior to that of the indigenous stock. However the introduced stock may not adapt well and performance then falls below that expected in the original environments. It is not possible to attribute such differences to climatic factors with any degree of certainty, however, since nutrition and management standards are never equivalent.

Information concerning climatic limitations on production can be obtained from experiments designed to ameliorate some aspect of natural climate. In cold regions winter housing may be practiced, especially when ground snow cover necessitates some form of hand feeding. In-wintering also aids efficient utilization of summer pastures and allows control of nutrition and of lamb losses during parturition. Different types of shelter may differ in the protection afforded. Czernek et al , (1960), Mikus (1964) and Stojanov (1967) compared part-open sheds which protect against rain and direct wind, with traditional closed barn housing for wintering of ewes. No significant differences were found, although Czernek et al (1960) reported better food conversion ratios for lambs born in half-open sheds. Madison (1966) found that ultraviolet irradiation of winter-housed Romanov lambs marginally increased live weights. Scottish Blackface sheep showed no consistent effects of winter housing in live weight change of pregnant or non-pregnant ewes at various levels of controlled nutrient intake (Armstrong et al., 1959; Doney, 1963; Russel et al., 1970). Similar work on Welsh Mountain ewes (Winfield et al., 1968; Robinson et al., 1970) has shown no consistent advantages due to housing. Analysis of blood plasma free fatty acid or ketone levels have also been used either to assess the nutrients required in different environments (Russel et al., 1967), or to measure responses to fixed levels of feed (Russel et al., 1969; Robinson et al., 1970). In terms of food requirements the results have ranged from around 10% in favour of the outdoor sheep to around 15% in favour of the indoor groups. Thus the additional energy cost of failure to protect from the winter climate is small compared with the cost of housing. However, to some extent the response is dependent on the genotype. Doney and Russel (1968) found that Merino ewes had higher winter maintenance requirements than either Scottish Blackface or Romney ewes when individually penned outdoors.

Cold also affects appetite. In the short-term, cold stress may reduce voluntary food intake, for example after shearing in cold weather (Webster et al , 1966) and after acute cold stress in climate chambers (Slee et al., unpublished). In the long-term, however, energy intake generally increases, providing suitable food is available. For example Winfield et al. (1968) recorded an increase due to winter exposure and Wodzicka-Tomaszewska (1963) reported a large increase during three months after shearing.

At the other climatic extreme, Marshall and Dixon (1969) found no advantage in housing Merino ewes in Western Australian conditions whilst Rea et al., (1963) and Arehart et al. (1969) found no significant effect of high shed temperatures on lamb growth rate, despite indications that shorn lambs were superior at 23°C and unshorn lambs at 7°C. Similarly Cockrem (1967) found that unshorn lambs grew faster at 7°C than at 18°C. Quartermain (1964) showed that summer ambient temperatures in S. Rhodesia (26-32°C) had little effect on intake or other responses if some protection from solar radiation was provided.

Although heat stress may cause decreased appetite (Alexander et al., 1971), lower food intakes can improve heat tolerance by reducing metabolic rate and the consequent heat load (Graham, 1964). Also Shelton and Huston (1968) showed that high food and protein intakes might be deleterious with respect to female reproductive performance in hot conditions Decreased food intakes may offer a further advantage when the energy cost of obtaining food is high. For example, in the study of Shafie and Sharafeldin (1965) breed differences in heat tolerance were partly related to behaviour in that the least-tolerant Texel continued to graze in the sun until exhausted. Burns (1968) found that bush grazing, even in indigenous sheep, caused increases in metabolic rate.

5. REPRODUCTION

The influence of climate may be considered in terms of the phase in the reproductive cycle at which climatic stress is applied. The onset of oestrus in breeds with seasonal breeding capacity may be affected by mean air temperature in the preceding summer period as well as more directly by the day length cycle. Thus Lees (1966) and Neville and Neathery (1970) found that longer oestrous periods followed higher summer temperatures, Tyagi and Lavania (1969) found earlier onset of breeding in indigenous Indian sheep compared with crosses to the Polwarth breed. Te Brake (1969) reviewed the effects of climate and other factors on the onset of oestrus. Baihanov (1965) found that the duration of oestrus was affected by different diurnal temperature ranges such that the higher the temperature range the lower was the mean length of the oestrous period.

There are no recent reports on the effect of heat on ovulation rate, although Lokvancic (1964) reported the mean number of corpora lutea to be greatest in intermediate conditions of temperature and humidity. In a cool climate Griffiths, Gunn and Doney (1970) found that intermittent exposure to rain during the oestrous cycle reduced ovulation rate.

Heat has important effects on early embryonic mortality. For example the studies of Dutt (1963, 1964); Woody and Ulberg (1964); Smith et al. (1966); Thwaites (1967a, b; 1970) show high mortality rates in heat stressed ewes and suggest that the most critical period is in the early cleavage phase of embryonic development. The effect is not mitigated by age of ewe (Thwaites, 1968), and is dependent on the elevation of body temperature. In field or experimental conditions where high temperatures were not continuous the effect was reduced (Smith, 1965; Thwaites, 1969; Entwistle, 1970; Rich et al., 1970). Spies et al. (1965) found that wetting and shelter from direct sunlight reduced the embryonic mortality rate, but that additional exercise in the heat increased mortality. Treatments were again associated with different mean rectal temperatures. Cockrem and McDonald (1969) related rectal temperature in a grazing flock to implantation rate and found a negative correlation.

Griffiths et al. (1970) found no significant effect of cold stress on implantation failure although some interaction with body condition was indicated. The relative importance of all causes of pre-natal mortality in sheep has been reviewed by Edey (1969).

Exposure to high ambient temperature at later stages of pregnancy may cause death or foetal dwarfing (Yeates, 1958; Ryle et al., 1961; Alexander et al., 1966; Moule, 1966; Shelton et al., 1968; Alexander et al., 1971). Influence of cold in late pregnancy has not been detected, but it might increase nutrient requirements when ewes are usually undernourished.

Male reproductive efficiency can be affected by heat but probably not directly by cold. Experimental exposure to high temperatures impairs sexual activity (Lindsay, 1969)and various characteristics of semen quality, including motility, concentration, and percentage of abnormal sperm (Brooks et al., 1962; Moule et al., 1963; Djanuar, 1965; Fowler et al., 1966; Rathore et al., 1967; Rathore, 1968; Howarth, 1969; Lindsay 1969). These effects relate to testis temperature (Moule et al , 1963; Makevnin, 1963; Fowler, 1968b; Fowler et al., 1968; Rathore, 1970). In field situations with fluctuating temperatures, similar, but less severe, effects have been noted (Malikov, 1963; Lamont, 1964; Loggins et al., 1964; Sahni et al., 1967). Provision of shade may alleviate the effects further (Makevnin, 1963). Conversely Steynberg and Hugo (1964) found no effect on semen quality of seasonal temperature, shade or breed of ram.

In general sheep in fleece seem well protected against a wide range of ambient temperature fluctuation. The adult sheep is probably reasonably comfortable within the range $-5^\circ C$ to $+20^\circ C$ in still air with minimal radiative exchange. Important effects on productivity are unlikely to occur unless the effective temperature range is widened by more than $20^\circ C$ at the lower end or $10^\circ C$ at the upper end. However genetic differences will modify these estimates. In contrast, the closely shorn sheep may only be comfortable within a narrow zone varying from about $+22^\circ C$ to $+32^\circ C$ and will suffer some degree of stress when the upper and lower limits are extended by more than $10^\circ C$ or $5^\circ C$ respectively.*

REFERENCES: see pp. 134–139.

* We gratefully acknowledge the cooperation of the Commonwealth Bureau of Animal Breeding during the preparation of this manuscript.

Meteorological Effects on Wool Production of Sheep

by

J.C.D. Hutchinson

1. INTRODUCTION

This paper deals with the effect of ambient temperature on wool growth, the circannian* rhythm of wool growth, the effect of climate on the quality of wool and man's influence on the climate of wool-producing sheep. Earlier work on climate and wool growth has been briefly reviewed by Hutchinson and Wodzicka-Tomaszewska (1961) and Hutchinson (1965) and that on wool production of sheep in the field by Brown and Williams (1970). When such work is discussed here the references of the review articles only are given in most cases.

2. WOOL GROWTH

From the field there is a survey covering the whole of Australia, a country large enough to have variety of climate but in which the general system of animal husbandry is fairly uniform and the genotypes are well known (Brown et al., 1970). Productivity per head decreases from South to North. In the coastal districts locality is confounded with breed, but in the semi-arid zone the sheep are Merinos with, however, a concentration of the high producing South Australian strong wool strain in the north of South Australia and parts of New South Wales. The authors thought that the gradients of productivity might be the consequence of local variation in availability of feed rather than the direct result of climate. Climates of the localities considered are described in "The Australian Environment" (Leeper, 1970).

Thwaites (1967, 1968) has shown in laboratory experiments that a hot environment can decrease wool growth in Merino sheep but the decrease can be accounted for by decrease in feed consumption. However, there is a possibility, not yet proved, that heat during gestation may affect follicular development in the foetus so that wool production in the adult is reduced (Alexander et al., 1971). Cold does not appear to affect wool growth provided that the site where it is measured is protected from cold, for example by the fleece (Bennett et al., 1962; Downes and Hutchinson, 1969); but if the site is not protected, wool growth is decreased and

* Term used by Pengelley (1967) to describe a rhythm which, when it is not entrained to environmental influences, has a period of about a year.

the decrease may be large (Doney et al., 1967b; Wodzicka-Tomaszewska et al., 1968; Downes et al., 1969; Lyne et al., 1970; Jolly et al., 1970).

On a constant ration there is an annual rhythm of wool growth. (Bennett et al., 1962). In all breeds examined, except the Bikaneri, maximum growth is in summer and minimum in winter (Hutchinson et al., 1961; Hutchinson, 1965). In Bikaneri ewes a reversed rhythm has been observed (Nagarcenkar et al., 1964 a). The rhythm is small in most Merino strains and larger in other breeds examined (Hutchinson et al., 1961; Hutchinson, 1965; Doney, 1966; Hutchinson et al., in preparation b). Several investigations have shown that it can be entrained by the photoperiodic rhythm (Hutchinson, 1965; Hutchinson et al., in preparation b). The rhythm has strong endogenous properties in that it can continue without loss of amplitude or timing for at least 2½ years in an equatorial daylength with exaggerated reversed seasons of temperature and humidity (Hutchinson, 1965; Hutchinson et al., in preparation b). Hart and Wildman in earlier experiments showed that, when temperature and humidity are uncontrolled, the rhythm will continue for at least a year with a variety of fixed daylengths (Hutchinson, 1965). In these experiments there was some evidence that short daylengths increased wool growth, particularly in winter and spring. Nagarcenkar and Bhattacharya (1964 b) observed a small increase in wool growth of Bikaneri ewes with a fixed long daylength as well as a larger increase with a fixed short daylength. Hart found that when Corriedale ewes were kept in complete darkness by covering their heads and eyes with hoods the rhythm did not continue for more than a year (Hutchinson, 1965). Apparently, continuous darkness affects the rhythm in a different way from a fixed daylength.

It has been suggested that the circannian rhythm of wool growth may be a modification of an archaic pattern of shedding and regrowth in which the shedding phase has been lost (Hutchinson et al., 1961; Hutchinson, 1965). The rhythm of shedding and regrowth of the kemp fibres of the Limousine sheep is entrained by daylength (Rougeot, 1957, 1961). In Wiltshire Horn lambs, born and reared in continuous dim light, the first season of shedding is protracted and incomplete, as if photoperiodic change or brighter illumination were required to initiate the annual rhythm of shedding and regrowth (Slee, 1965). Symington found that Persian Blackhead rams did not shed their coats when they were housed for eight months (Hutchinson, 1965).

When a seven-month cycle of ambient temperature is applied to Merino ewes and wool growth is measured on patches of skin unprotected from the cold, a rhythm of wool growth is generated that has a much larger amplitude than the circannian rhythm (Hutchinson et al., in preparation a). It is also much larger than the rhythm exhibited when the same treatment is applied with a period of one year Apparently, there is a natural semestrial periodicity. Since the effect of cold is on the follicles only, the semestrial periodicity may be local. When the circannian rhythm is entrained to a seven-month protoperiodic year amplitude is not larger than normal (Hutchinson et al., in preparation b). The physiological mechanism of the circannian rhythm is unknown. Presumably it must have a component in the brain because hooding the head abolished the rhythm in Corriedale ewes (Hutchinson, 1965).

In the field the seasonal variations of wool production are different from those on a constant ration. In Merinos the effect of nutrition overrides that of the weak circannian rhythm and wool growth depends mainly on the green feed available (Brown et al., 1970). Consequently, in a mediterranean climate the maximum production is in winter and spring and the minimum in summer (Hutchinson, K.J. 1962). On the other hand in the Cheviot and Scottish Blackface, and perhaps some other breeds, nutrition in winter has little effect on wool growth; at this season production appears to be controlled by the circannian rhythm. In summer wool growth is sensitive to nutrition (Doney, 1964, 1966; Doney et al., 1967 a). Thus, the circannian rhythm has no influence on productivity of Merino sheep, but does influence productivity of some other breeds by limiting winter production.

3. WOOL QUALITY

Ultraviolet light degrades the tips of the wool fibres so that they may break off during processing. This decreases the value of the wool. Seeds, other vegetable matter and dirt get into the wool in quantities depending on the locality and therefore on the climate. Wool with more than around 6% of vegetable fault cannot usually be used for topmaking. In some districts of Australia, which in other respects are suitable for fine wool production, 12-15% of vegetable fault is not uncommon.

"Fleece-rot", a bacterial dermatitis with exudation, matts the wool fibres and is usually accompanied by staining of the wool. The cause is presence of moisture on the skin over several weeks (Hayman, 1953) and it has been associated with particular patterns of rainfall (Hayman, 1955). The relation between Merino fleece character and susceptibility is described by Dunlop and Hayman (1958). In India's "Canary colouration" of the wool, in this case without fleece-rot, is a serious economic problem. In some districts sheep have to be shorn twice a year to separate wool grown in the wet season from that grown in the dry season. Consequently, there have been many recent investigations in India on the influence of climate and fleece type on incidence of staining (Nagarcenkar et al., 1964c, d; Narayan et al., 1968, which includes other recent references; Narayan et al., 1969).

"Fly-strike" is caused by the development of blow fly maggots under the skin and outbreaks can be severe after wet weather. It is most common in the neighbourhood of the tail, anus and vulva, but if fleece-rot is present there may be fly-strike under the fleece which reduces the value of the wool.

Mycotic dermatitis, an infection of the wool follicles by *Actinomyces dermato-*

nomus, is associated with high rainfall, like fly-strike, and sometimes responsible for economic loss.

4. THE EFFECT OF SHEARING SHEEP ON THEIR THERMOREGULATION

Man, by shearing sheep, disturbs their heat regulation. The fleece protects the animals from the heat of solar radiation as well as from cold (Hutchinson et al., 1961; Robinson, 1969). After shearing, sheep are often lost from cold but not from heat. The loss in Australia averages less than 1%, but can occasionally be catastrophic in localised areas in both Australia and New Zealand (Geytenbeek, 1963; Hutchinson, 1968). The heavy losses occur only in a combination of heavy rain and strong wind. Cheap polyethylene coats to protect recently shorn sheep have been developed (Panaretto et al., 1968; Hutchinson et al., 1969, 1970).

Since Roman times attempts have been made to protect the wool of sheep with coats. However, in the past these attempts have not been successful because the materials have been too costly and their durability has been inadequate. The advent of new cheap plastic materials and the continued competition of synthetic fibres with wool has renewed interest in the problem of protecting the fleece (Lipson et al., 1970). It is possible that the coats now being designed may reduce fleece-rot and fly-strike as well as excluding contaminants and preventing damage from ultraviolet radiation.

REFERENCES: see pp. 139–140.

Meteorological Effects on Pigs

by

D.L. Ingram

1. INTRODUCTION

The effects of meteorological conditions on farm animals can be assessed in terms of both agricultural productivity, and basic physiology. In practice, however, productivity depends on a number of variables such as quantity and quality of food intake, disease, and management, as well als climate; moreover, climatic conditions may have an indirect effect on these other factors. For this reason, and to facilitate comparison with man and other animals, an appreciation of the effects of climate on the pig is probably best achieved by concentrating attention on the animal's physiology. The subject has received comprehensive treatment by Mount (1968), and also by others in a series of articles in a special issue of Monatshefte für Veterinarmedizin on temperature regulation in pigs (Lyhs, 1967; Holub, 1967; Herbst, 1967; Leucht, 1967; Nichelmann, 1967; Studzinski et al, 1967; Schülke et al., 1967; Adam, 1967); work on the new-born pig has recently been reviewed by Curtis (1970).

2. INFLUENCE OF HOT CLIMATES UNDER LABORATORY CONDITIONS

The pig's reaction to high temperature has been investigated under laboratory conditions by Ingram (1964a, b), Roller and Goldman (1969) and Lyhs et al., (1969) and its tolerance of heat has been found to be relatively poor. Although pigs have glands in the skin which have the appearance of sweat glands and which can be made to discharge after the administration of adrenaline, the animals do not appear to sweat (Mount, 1962; Ingram, 1964, 1967). The only other channel of evaporative heat loss is via the respiratory tract and even here the pig's capacity to increase its respiratory ventilation rate falls short of that of other farm animals (Morrison et al., 1967; Ingram, 1970). The consequences of this limited ability to lose heat by the evaporation of water are that pigs tolerate heat poorly, but are less affected by humidity than some other large mammals (Ingram, 1965). If heat tolerance, as judged by the rate of rise of rectal temperature, is plotted against an "effective temperature" scale developed for man then points obtained at high and low humidities fall on different lines, and the fit is only partially improved when the "effective temperature" scale suitable for cattle is used; the "effective temperature" scale for pigs takes relatively little account of humidity (Fig. 1). The effects of humidity and high D.B. (Dry Bulb) temperatures on gain in body weight has been

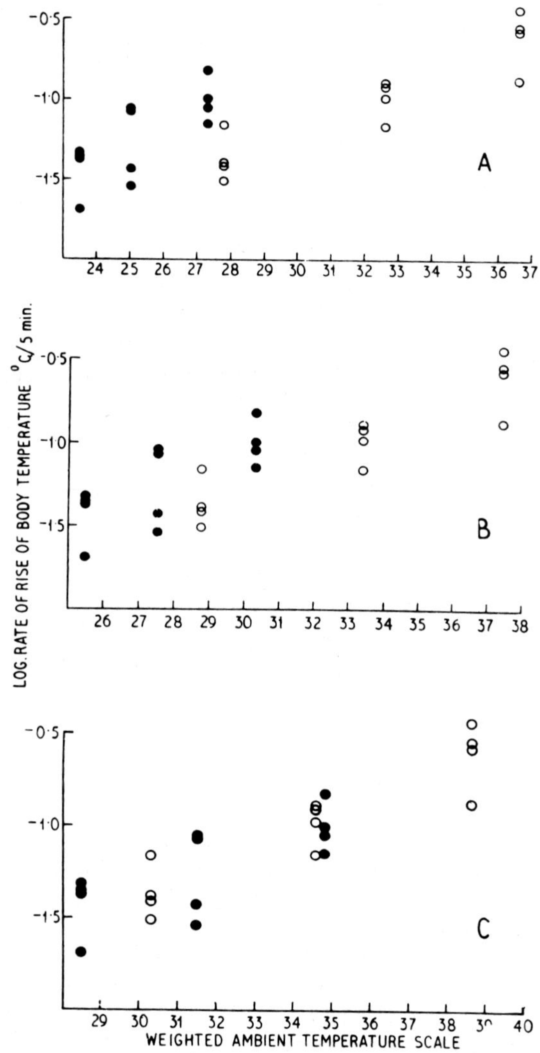

The relationship between log rate of rise of body temperature and 3 weighted environmental temperature scales, namely:
Fig. 2A: DB × 0·15 + WB × 0·85.
Fig. 2B: DB × 0·35 + WB × 0·65.
Fig. 2C: DB × 0·65 + WB × 0·35.
● = low humidity.
○ = high humidity.

Fig. 1. Effective temperature scales combining Wet and Dry Bulb temperatures plated against the logarithm of the rate of rise of rectal temperature in the pig. A is the weighting which is correct for man, B for cattle, and C is that which proved best for pigs (From Ingram (1965) by permission).

investigated by Pfeiffer (1967) and Heitman et al. (1958) and a theoretical relation from which gain in weight can be predicted has been published by Morrison, et al. (1968). Under climatic conditions where the D.B.-temperature exceeds 40°C pigs often die in hyperthermia, but the provision of a shower or mud wallow enables the animal to make up for its limited ability to lose body water by evaporation (Heitman et al., 1962). This evaporation of water from a layer of wet mud provides cooling similar to that from a sweating man and sufficient water may be held in the mud to sustain a high rate of heat-loss for 2 hrs (Ingram, 1965). Moreover, since the pig has very little hair, the heat of evaporation is likely to be derived from the body surface to a greater extent than in animals with a thick coat.

Some slight increase in the ability to tolerate high temperatures may result from the decrease in resting metabolism which occurs after exposure to high temperatures for 10 days (Ingram et al., 1965). This adaptation appears to depend on a decrease in thyroid activity since thyroxine output declines when young pigs are kept at a high D.B. temperature, but no decline in metabolism was observed in thyroidectomized pigs on a standard replacement dose of thyroxine after 3, 5 and 7 weeks exposure to a high D.B. temperature (Ingram et al., 1964). When pigs were raised in a hot room on a restricted diet they developed longer legs, tail and ears and had a better blood supply to the skin than controls brought up in the cold (Fuller, 1965; Weaver et al., 1969; Ingram et al., 1969) and these modifications may well be adaptive under natural conditions. Reproduction is adversely affected by high temperatures the incidence of anoestrous periods being increased and the percentage of gilts becoming pregnant being decreased (Teague et al., 1968).

3. INFLUENCE OF COLD CLIMATES UNDER LABORATORY CONDITIONS

Although the pig has a sparse hair coat it has nevertheless been kept successfully in Alaska where it has been studied by Irving (1964). Under these conditions the critical temperature was below 0°C in a 50kg boar, but during the winter when the same animal weighed 150 kg the critical temperature was much lower, and thermal equilibrium was still possible at a D.B. temperature of –30°C. At least part of this tolerance to cold was related to improved insulation achieved by lowering the skin temperature so that at –30°C D.B. the surface of the skin was 5°C. According to Irving (1964) the decrease in heat loss brought about by the improved insulation enables pigs to be kept during the winter at very little extra cost in food intake. Sørensen (1962) has shown that exposure to cold results in the deposition of extra back fat and MacGrath et al. (1968) found an increase in the unsaturated fatty acid content of back fat at low ambient temperatures. Exposure to severe cold during growth results in the development of small ears and limbs, and an increase in hair density (Fuller, 1965; Weaver et al., 1969) all of which probably helps to reduce heat loss.

Behavioural and postural changes may also assist the pig to conserve heat under cold conditions. Changes in posture alter the proportion of the total body surface available for heat exchange by radiation and convection. Mount (1964, 1968) has carried out a series of experiments on new-born pigs exposed in chambers where the radiant and air temperatures could be altered independently. He estimated that the change in posture which occurred when the ambient temperature was reduced from 30°C to 20°C was associated with a decrease in the radiating area from 76% to 67% of the total body surface. At the same time the effective convective area decreased from 86% to 76% of the total. Conductive heat loss is also influenced by posture for while the pig may lie down on its side with maximum contact of the body with the floor under warm conditions, when it is cold it tends to lie down on the lower part of its limbs and so reduce heat loss (Mount, 1967). Estimates of heat loss to the floor by pigs and evaluation of the use of different types of floor material for the purposes of pig husbandry have been made by Kelly et al. (1964), MacHardy (1960), Harman et al. (1968), Mount (1967) and Stephens and Start (1970), and the evidence suggests that the type of floor used may be of significance under both warm and cold conditions. A further mechanism for combating cold conditions available to a group of animals is to huddle together. In both new-born and growing pigs this type of behaviour reduces the metabolic effort required to maintain body temperature and enables a group of pigs to tolerate a colder environment than a single animal (Mount, 1960; Holmes and Mount, 1967), the group tending to behave like a single pig of the same total weight. Similarly an improvement in the tolerance of low temperatures by baby pigs as the result of providing straw bedding has been measured by Stephens (1971).

4. INFLUENCE OF FLUCTUATING TEMPERATURES

Pigs are frequently exposed to temperatures which fluctuate about a mean such as occurs naturally in the course of 24 hrs and several studies have been made in which the effects of a constant ambient temperature have been compared with those produced by cycling conditions. Under conditions of constant temperature the actual deep body temperature of the pig varies very little but in pigs living out-doors the body temperature falls during the night (Ingram and Legge, 1970). Heat loss, however, was found by Cairnie and Pullar (1959) to vary cyclically even under constant lighting and temperature conditions although the variation did become more obvious when the pig was also subjected to a dark-light cycle. Responses of pigs to fluctuating temperatures on the one hand, compared with a steady temperature equal to the mean of the fluctuating temperature on the other have been made by Bond et al. (1963, 1967). They found that physiological variables such as skin temperature tended to lag behind the change in environmental conditions as might be expected, and although the total heat loss over 24 hrs was the same under both fluctuating and constant conditions the channels of heat loss differed because of the increase in evaporative loss and decrease in convective loss as the D.B. temperature increased above the mean. Sϕrensen (1962) concluded that as judged by weight gain a fluctuating temperature corresponded to a constant temperature close to the mean value of the cycle.

5. INFLUENCE OF AIR MOVEMENT

Besides the actual D.B. temperature another meteorological variable of particular significance in the cold is the amount of air movement. Measurements made with the aid of a heat flow disc and a hot wire anemometer (Mount et al. 1964) demonstrated that heat loss from a localized region on the back of a three month old pig was approximately doubled as wind-speed increased from still air to 100 cm/sec at 15°C, while at higher D.B. temperatures there was rather less effect (Fig. 2). Age was also found to be an important factor, the younger pigs being more affected by air movement. Heat loss from the whole body of new-born pigs has also been related to air movement by Mount (1966) who found that although increased air movement is associated with an increase in oxygen consumption (and by implication heat loss), the increase is not as great as would be expected from measurements made on a model. Mount believed that the difference was accounted for partly by the effects of change in posture by the animal in response to higher air movement, partly by the lowering of skin temperature as a consequence of vasoconstriction, and partly because air movement about the pig was not as uniform as that round the model. An important aspect of the effect of air movement on pigs is that small changes in the range 0-100cm/sec have a marked affect on heat loss because of the sparse hair coat. In this connection Mount (1968) found that the disturbance of the boundary layer of air round the new-born pig caused by shivering leads to an increase in convective heat loss.

6. INFLUENCE OF CLIMATE UNDER NATURAL FARM CONDITIONS

Under laboratory conditions where each variable in the environment can be closely controlled the separate effects of each component can be assessed; but under practical farming conditions there is considerable interaction not only between meteorological factors but also with other variables such as food intake, management and disease. The difficulty of measuring factors such as air movement and of making comparisons between different types of pig houses has been stressed by Gordon (1962), but the importance of wind velocity is nevertheless underlined by such investigations as those of Gunnarson et al. (1967) who found it to be the only variable which significantly affected food intake under the conditions of their study. In cold weather heating pig houses may improve food conversion, but the advantage decreases in older pigs (Jensen et al., 1969) and in some houses outdoor conditions still affect the climate inside in spite of heating (Seman, 1967, 1968). At temperatures in the region of 10°C fattening pigs need more food per unit of growth than at 15°C to 20°C, but at 30°C although food intake may be reduced, so also is growth rate (Seymour et al., 1964; Holme et al., 1967). Tonks and Smith (1965) found no significant differences in pigs raised at 21°C and 28°C when the animals were between 20 and 70 kg, but between 70 and 90 kg the pigs at 21°C had a 10% higher growth rate. In a study of growth of pigs during the first 10 days of life, Boaz and Walters (1966) found no difference in growth rate between 7°C and

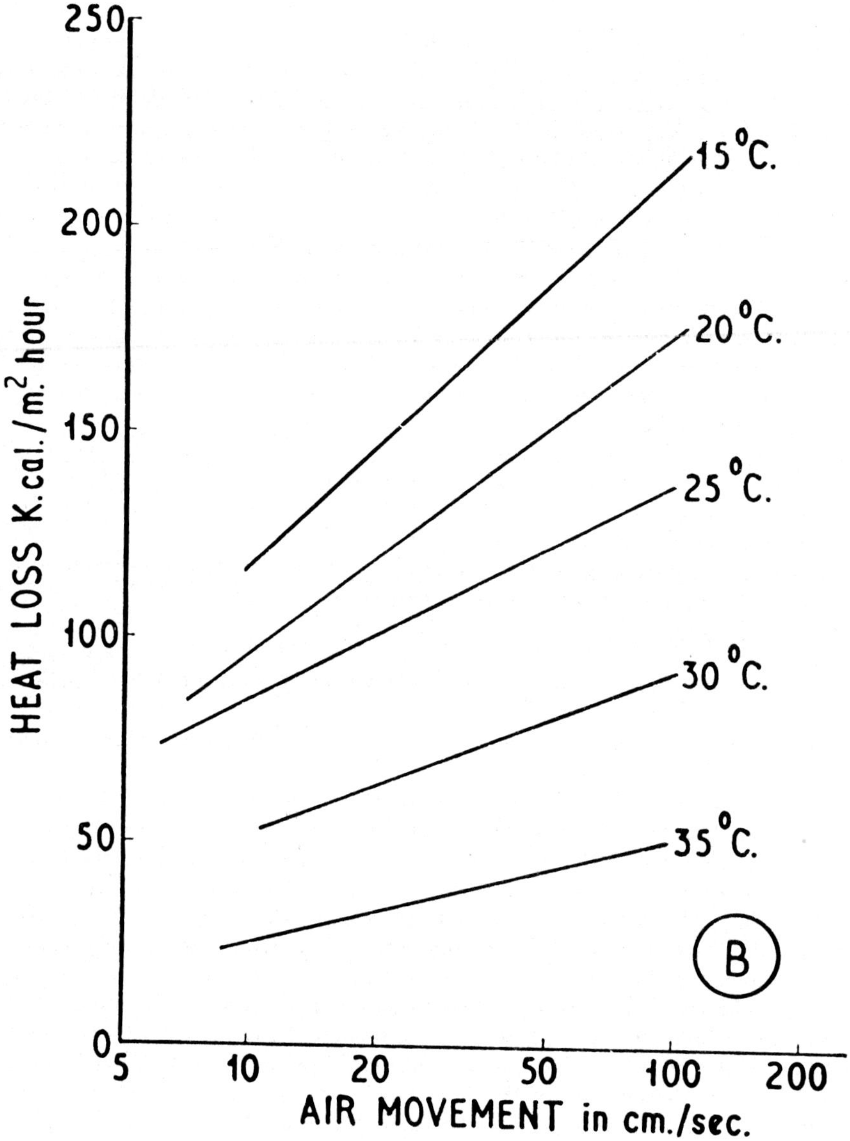

Fig. 2. Effect of air movement on heat loss from the trunk of a three month old pig (From Mount and Ingram (1964) by permission).

18°C D.B., but milk intake was 15% higher for animals kept at the colder temperature. Given plenty of straw pigs will modify the environment in such a way that houses of very different thermal insulations have little effect on growth rate, or food conversion (Cunningham et al., 1963). A similar adaptability by groups of pigs kept at 9°, 20° or 30°C was also demonstrated by Holmes and Mount (1967) using a pen calorimeter which was constructed to resemble the type of pig unit used in farm practice.

7. ANALYSIS OF THE BEHAVIOUR OF PIGS IN RELATION TO THERMAL STRESS

The observation that pigs can modify the effect of the climate by means of their behaviour has led to studies on the extent to which pigs can detect changes in ambient conditions (Baldwin et al., 1968 a, b). In these experiments a pig aged 10 weeks was placed into a cage 90 cm x 60 cm and eight 250 w infrared heaters were suspended 20 cm above the pig's back. At one end of the cage a metal box with a panel 8 cm square in the centre was attached to the wall. If the pig pushed the panel with its snout the heaters came on for 3 sec and then went off. Pigs exposed to 5°C in this cage soon learned to operate the switch and in a series of experiments it was found that below 25°C the rate at which the heaters were switched on increased with a fall in D.B. temperature (Baldwin et al , 1967). These findings agree well with the estimate of 25°C for the critical temperature in pigs of this age (Ingram, 1964). In further studies it was shown that when a pig was able to operate the heaters and so modify the environment, then its oxygen consumption remained minimal over a range of D.B. temperature below 25°C; but when no heaters were provided a fall in D.B. temperature from 25°C to 10°C was accompanied by a doubling of oxygen consumption. Confirmation of the pig's ability to manipulate the ambient conditions in a quantitatively appropriate way was obtained in other experiments (Baldwin et al., 1967) in which it was shown that the pigs would turn the heaters on twice as often for a 6 sec period of heat as they would for a 12 sec period. In another study where the conditions were less closely controlled a group of pigs were housed in a hut with an outside yard. A bank of heaters and a switch were placed in the yard and the pigs soon learned to obtain extra heating on cold days, but they went into the hut to sleep at night. By contrast a single pig kept in a cage in a temperature controlled room would operate the switch over a 24-hrs period (Baldwin et al., 1967). When the pigs were kept in a grass paddock and the heaters were placed in one corner, additional heat was only rarely sought even on cold days.

Pigs also learned to operate a panel switch in order to turn off a draught of air blown into the cage by a fan and again the rate at which the fan was turned off increased with a fall in D.B. temperature. The D.B. temperature below which the switch was operated was in this instance 30°C and corresponds to the higher critical temperature expected under conditions of high air movement. It was believed, however, that a draught may contain aversive components in addition to those associated with heat loss since the time for which the fan remained off did not influence the rate at which the switch was operated in the manner which would have been expected if thermal comfort was the only factor involved.

8. EFFECTS OF WEATHER ON THE BEHAVIOUR OF PIGS

The above experiments leave little doubt about the pig's ability to judge the climatic conditions and make appropriate behavioural responses and are in agreement with the findings under husbandry conditions that pigs may grow equally well over a range of D.B. temperatures. The next question which arises is the extent to which meteorological variables influence behaviour under near natural conditions when pigs have the opportunity to make a wide range of responses. In a hot climate pigs are known to make use of artificial cooling (Heitman et al., 1962) when they are kept under farm conditions, but few observations have been made on pigs on a temperate climate. In this laboratory the behaviour of groups of four pigs aged 8 to 14 weeks was studied under natural conditions and meteorological variables were recorded at the same time (Ingram and Legge, 1970). The animals were observed on a patch of land approximately 60 m x 50 m which consisted of a belt of trees 30 m wide with a grass paddock to the S.W. and another to the N.E. A pig-hut provided with straw bedding was placed within the wooded area and the pigs were fed once a day an amount of food equal to 5% of body weight.

During the daytime in dry weather the pigs did not begin to shelter in the hut until the D.B. temperature was below 5°C. This temperature is well below the critical temperature for pigs of this age as measured in the laboratory and it was concluded that the time spent outside involved the animals in some degree of thermal discomfort. One of the factors which encouraged the pigs to remain outside was the opportunity it offered to root for more food, since when animals were fed ad libitum the time spent in the hut increased. During rain or snow the animals spent considerably more time in the hut, but provided the D.B. was over 5°C some time was spent outside even while it was raining in spite of the fact that this probably involved them in a high degree of thermal discomfort. In both summer and winter the animals made visits to the hut with increasing frequency from about 6.00 hr onwards and from 9.00 hr to 16.00 hr the animals usually made only one visit outside. From 16.00 hr or 17.00 hr the animals' visits outside became more frequent until about 21.00 hr when they usually remained outside for most of the time.

Choice of position within the enclosure seemed to depend chiefly on air movement. Thus with a prevailing S.W. wind the animals spent most of the time in the S.E. paddock and further analysis showed that the proportion of time spent in the wood increased in proportion to the shelter it provided when the wind direction changed. If access to the S.E. paddock was prevented by closing a gate the pigs spent most of the time in the woods rather than go into the more windy S.W. paddock, but if the wind direction changed making the wood the most windy place the proportion of time spent in the S.W. paddock increased (Fig. 3). This tendency to avoid areas of high movement agrees well with both laboratory and farm findings, but as mentioned before it may be that wind is avoided for reasons in addition to those connected with heat loss.

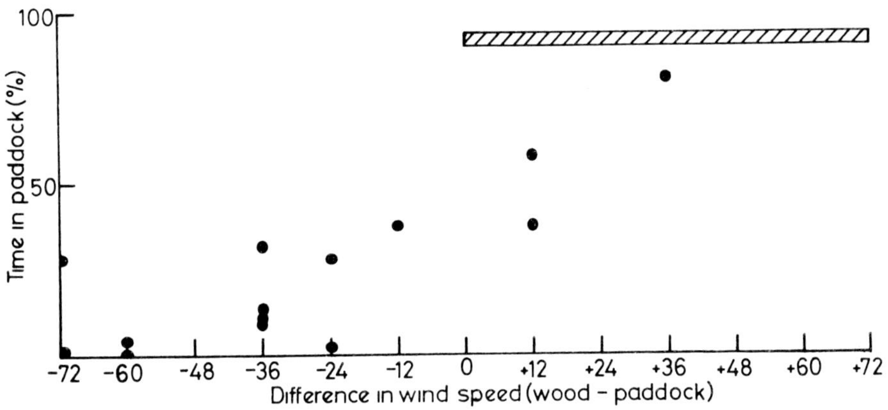

Fig. 3. Effect of air movement on time pigs spent in an open paddock or a wooded area (From Ingram and Legge (1970) by permission).

The mean radiant temperature appeared to have no effect on the pigs choice of environment even after allowance had been made for differences in air movement. This finding contrasts strongly with the readiness to perform an operant response for infra-red heat and probably reflects the additional behaviour patterns the pigs may indulge in when they are kept under natural conditions.

These investigations have lead to further studies in which the heat loss from pigs living outside is estimated by means of a heat flow disc the output of which is transmitted by radiotelemetry (Heal et al., 1970). Tentative figures for the mean heat flow from round the trunk of a three month old pig in an open paddock are about 50 $Kcal/m^2$ hr in summer to about 160 $Kcal/m^2$ hr under fairly mild winter conditions in England. Studies of this kind using telemetry techniques will, it is hoped, provide us with a better understanding of the effects of climate and weather on domestic and other animals.

REFERENCES: see pp. 140–143.

Meteorological Effects on Poultry With Special Reference to Disease, Egg Production and Quality

by

W.O. Wilson

Poultry has served man in many roles – as a provider of nutritious and economical food, warmth, entertainment, companionship and as a laboratory animal for many kinds of research. This review covers only a small area of the latter that deals with biometeorology research reported in the past seven years.

1. POULTRY DISEASES

At present the most serious disease of chickens is Marek's disease. The natural environment of the chicken harbors the infectious agent of Marek's disease. The herpesvirus agent has been detected in feces, saliva and dander from infected chickens according to a review by Witter (1970). The airborne particulate matter has been shown to be highly infectious. This route may account for the extreme contagiousness of the disease. It now seems likely that the infectivity present in the environment consists of complete virus associated with or originating from desquamated epithelial cells from the feather follicle since this site was found to be uniquely permissive for replication of the virus. Feather sheath cells are present in large quantities in the poultry house environment and are suspected of being responsible for airborne transmission of the virus. The employment of biologically filtered air under positive pressure has been successful in preventing the introduction of airborne virus diseases of poultry (Drury et al., 1969).

Concentrations of ammonia of over 50 ppm causes keratoconjunctivitis (Valentine, 1964). She found that levels over 20 ppm were not common in the summer in British poultry houses. However, Anderson et al. (1964) reported that levels as high as 110 ppm has been reported in poultry operations where ventilation rates were minimal. Further, he showed that a 20 ppm level caused visible damage to the respiratory tract of chickens and after three days exposure resulted in greater susceptibility to Newcastle disease virus in chickens and air saculitis in turkeys (Anderson et al., 1968). To relate this to a subjective measurement, man, they report, may detect levels as low as 5 to 10 ppm, while 50 ppm causes discomfort.

Based on a report by Walter (1966) once respiratory diseases are established in a house their spread by the air route is assured. His report indicated that the thres-

hold sanitation rate (the air change rate at which cross infection is avoided) for a man as 600-750 times the tidal volume. For chickens this was calculated by Charles (1970) to be 13 to 16 1/s or 24-34 cfm. These values are not practical to use even in the summer.
The health hazards of high litter moisture and associated high levels of ammonia in chicken houses are real.

Added humidity is one means of increasing the sedimentation rate of dust particles. Ackers et al., (1966) showed that survival of certain viruses was of shortest duration within the relative humidity range of 40 to 60% at 26°C, and longest at very high or very low humidity. If there is a case for regulating relative humidity in commercial practice it may be on the basis of reducing respiratory disease risk. In a poultry house with litter the concentrations of air borne particulate matter as high as 1.16 mg/ft^2 of air was reported by Anderson et al., (1966). They failed to find any effect of dust or mortality of chickens after exposure to aerosol Newcastle disease virus.

Condemnations of broiler carcasses are highest in the winter and lowest in summer months suggesting an environmental stress. *Mycoplasma gallisepticum* infection increases condemnations and especially when aggravated by stressor agents. Neither exposure to temperature extremes (Reece et al., 1969) nor ventilation rates of 0.5, 1.5 and 3.4 cfm/bird (Griffin et al., 1967) appear to affect the condemnation rate of broilers. The relative importance of the "triggering" stress factor, especially environmental, is not clearly understood, and is subject to considerable controversy.

The effects of air pollutants besides dust were reviewed by Lillie (1970). He concluded that poultry appears to be much less susceptible to air pollution than other farm animals. However, several pollutants have demonstrated toxic effects. Most of these are outgrowths of bacterial decomposition of poultry wastes. These organisms convert waste substances into ammonia, hydrogen sulfide, carbon dioxide and methane. Carbon dioxide and lead poisoning data for poultry are inconclusive or unreliable. Toxic effects on poultry of asphalt soot, carbon monoxide, hydrogen sulfide, nitrogen oxides, and dusts have been demonstrated in field studies (Lillie, 1969).

2. FACTORS, AFFECTING EGG SHELL QUALITY

In an exhaustive review of factors affecting egg shell quality Walford and Tanaka (1970) cited reports indicating that high temperatures had the most effect on this trait. Shell thickness may be experimentally affected by high CO^2 concentration. While ammonia at levels below 105 ppm is without effect. Thinner shells crack more easily are of concern to poultry egg producers.

Dirty egg shells result when the litter moisture is high. Winn and Godfrey (1967) found that above 80% relative humidity problems arose with wet litter. Space allotment per bird is as important as is relative humidity according to Deaton et al., (1969) who reported that when 465 cm^2 per bird (crowded) the mixing of the litter with the wet droppings was reduced. Heat and high humidity significantly decreased the thickness of the whole shell and of all its layers except the membranes (El-Boushy et al., 1968). Structure as well as ultrastructure were unfavorable affected by climatic stress. The cuticle degenerates, the mammillae are rounded at their bases, the calcareous deposits are loose, irregular, lack normal columnar orientation and develop cavities. The organic spongiosa matrix is of normal density but shows larger vesicles, some cavities and fewer and shorter fibrils which are not so well oriented (El-Boushy et al., 1968).

The number of double-yolked, thin shelled and eggs without shell membranes was related to early sexual maturity by Lacassagne and Jacquet (1965). The lighting regimen determined the age at sexual maturity.

There is an association between ambient temperature and the number of blood spots in eggs according to Wilson et al., (1964). Hens kept at cooler temperature 21°C laid more eggs containing blood spots than hens kept in a warmer temperature of 32°C.

Egg size tends to be increased by a decrease in temperature during the last three hours of the light phase according to Payne (1966).

3. INFLUENCE OF TEMPERATURE

It is difficult to separate temperature, humidity and air velocity or ventilation as all interact to affect the physiology of body temperature control. Hyperthermia in parts of the world, however, is a serious threat, especially when uncontrolled in some operations, since biologically the chicken is poorly adapted for heat dissipation. Within the laying house an ideal temperature is between 13° and 30°C. The day-old chick is sensitive to both hypothermia and hyperthermia, if subjected to extremes some will die and the growth rate of survivors will be retarded. From 21.2°C to 37.8°C densities of 9.3 and 6.5 dm^2 made no difference on body weight gain. There was a trend for better feed efficiency at higher temperatures, but mortality and condemnations were not significantly affected at the temperature and density levels studied. The temperature inside a broiler house can have a daily or weekly cycle within reasonable limits when the mean temperature of the cycle is equal to the constant temperature, without seriously affecting the overall performance (Griffin et al., 1970; Siegel et al., 1970).

At about 42°C deep body temperature, panting initiates increasing respiratory frequency and minute volume but amplitude and tidal volume decreases (Whittow, 1965). Heart rate increases, mean arterial pressure and vascular resistance lowers as body temperature increases from acute increases in ambient temperature (Linsley et al., 1964). Respiratory alkalosis in cocks developed as body temperature rose. Circulatory disturbances are postulated as the ultimate cause of death from hyperthermia (Linsley et al., 1964; Whittow, 1965).

Egg production was higher in a cyclic temperature condition from 13 to 32°C than in a constant 18°C (Payne, 1966). Egg size also was higher in the cyclic condition.

4. INFLUENCE OF AIR VELOCITY

The velocity of air is a factor in convective heat control. Chickens at environmental temperatures below body temperature and a velocity of 150 meters per minute, body temperature control are less dependent on respiratory evaporation than with a velocity of 15 meters per minute; however, at temperatures above body temperature high air velocity appears detrimental and survival times are reduced (Siegel and Drury, 1968; Siegel, 1969). Respiratory evaporation is extremely important as a means of heat loss at high environmental temperatures. Layers at high temperatures (32.2°C) lose three times as much moisture by respiration as they do at low temperatures (-1.1°C) (Howes, 1968).

5. INFLUENCE OF LIGHT

Light's influence on reproduction and growth has probably received more research attention in recent years than any other parameter in poultry husbandry. Natural light springtime photoperiods enhances egg production. Light penetrates the retina of the eye and ultimately causes the secretion of the gonadotropins necessary for gonadal function. The three types of light regimens that induce gonadal growth in photosensitive birds have been listed by Wilson and Siopes (1967) as long photoperiods, frequent photoperiods and conditioned cyclic gonadal response.

The AGE OF ONSET OF SEMEN PRODUCTION is inversely related to the photoperiod length during the growing phase of cockerels (Parker et al., 1964; 1965) but does not alter the fertilizing capacity of the semen.

With the Japanese quail gonadotropic potency was found to be highest in birds exposed to continuous light and lowest in those given 4 hrs of light (4L/20D). Lightning regimens of 12 hrs or less were essentially non-stimulatory by 5 weeks of age while those of 14 hrs or with intermittent light were (Tanaka et al., 1965). Photoperiodic control prior to the onset of egg production is essential for subsequent reproductive performance. Unlike chickens, turkeys become photorefractory to increasing light during the growing period but this condition can be overcome by restricting the daily photoperiod to 8 hrs for 4 weeks. Dim light or brownouts (0.86 lux or less) was as effective as total darkness as a preconditioning light regime for out-of-season egg production (Leighton et al., 1969). As in chickens, increment of light as compared to constant light during the reproductive period did not significantly improve egg production of turkeys (Wilson et al., 1967). Between 10 and 20 lux and a photoperiod of 15 hrs are recommended for breeder turkeys.

According to Morris and Owen (1966) stray light of 0.4 lux or less entering poultry houses which are supposed to be blacked out can be ignored for photoperiodic purposes, thus total darkness is not necessary. Light intensity for stimulation of feed intake is lower than for reproductive photoperiodic stimulation (Morris, 1966). Minimum light intensity required for maximum egg production is more than 2 lux but less than 10 lux (Morris and Owen, 1966). It is generally accepted that 5 lux causes stimulation of ovarian activity in chickens. From a literature review, Morris and Fox (1968) suggest that although the orange and red bands of the spectrum induce the greatest photoperiodic response, there is little reason to prefer red lighting to white in a poultry house or red paint on a lamp does not increase the red light but eliminates the blue and green components.

Noise, feeding regimens (Wilson et al., 1964) and fluctuations in environmental temperature (Payne et al., 1965) affect oviposition timing but are much less effective than light-dark cycles. Increasing photoperiods during reproduction have been compared to constant photoperiods (Noles et al., 1964; Lillie et al., 1965 and others) and have failed to significantly affect annual egg production. Increasing the photoperiod during sexual development results in depressed egg production. Decreasing the photoperiod during pullet rearing delays the onset of reproduction and results in increased egg size (Bornstein and Lev, 1969). Uniform, precise changes in the photoperiod are unnecessary, the prerequisite for delayed sexual maturity is that the ultimate desired photoperiod is attained (Harrison et al., 1965; Shutze, 1966; Bornstein et al., 1969). The overall effect of changes in photoperiod may depend upon the total difference between initial and final photoperiods, irrespective of the manner or rate in which these changes take place (Yamada et al , 1966; Bornstein et al., 1969)

6. ACCLIMATION

The ability of the hen to acclimate to environmental conditions has been again demonstrated by Long et al., (1966) at a latitude of about 40°N. They compared winter ventilation rates of 0.95 1/s (2 cfm) per bird and 2.8 1/s (6 cfm) and observed no effect on egg production. Summer ventilation rates of 2.8 1/s (6 cfm) and 8.5-9.5 1/s (18-20 cfm) were also without effect.

The role of acclimation in birds has until very recently masked the presence of non-shivering thermogenesis (El Halawani et al., 1970). Muscle electrical activity used as a measure of shivering was markedly reduced in frequency and amplitude in five months cold acclimated cocks and was nearly absent after nine months of acclimation.

Several important interactions exist around environmental factors and production traits. Egg production from hens fed marginal protein content may fluctuate with temperature (Bray et al., 1961) and with photostimulation (Bray, 1968). Higher dietary levels and intakes of protein were required despite a decrease in egg production. Similar interactions between temperature, food intake and protein requirement of the laying hen were reported by Reid et al., 1965 and Summers, 1965.

REFERENCES: see pp. 143–145.

Meteorological Effects on Fertility and Animal Reproduction

by

D. Smidt

1. INTRODUCTION

In either a positive or negative way nearly all species of animals could be regarded as "important to human life". This chapter will deal mainly with animals providing food for human consumption, e.g., domestic animals, and, to a lesser extent, wildlife.

Animals live under various ecological conditions. Wild animals adapt, for the most part, to the meteorological situation of their location and, thus, meteorological factors have little influence on fertility. Although this also applies to adapted domestic animals, their reproductive processes could be disturbed if they are kept in unfavourable climatic conditions.

Meteorological influences on reproduction and fertility can act directly or indirectly. In females, sexual development, manifestation of sexual rhythms, sexual behaviour, and reproductive processes from the time of ovulation to parturition of offspring are modified by meteorological factors. In males, sexual maturity, reproductive behaviour, spermatogenesis, and fertilizing capacity of sperm are similarly influenced.

2. DIRECT METEOROLOGICAL INFLUENCES

The more important direct meteorological factors affecting fertility are light, temperature, and humidity. Other factors include altitude, air pressure, air movement, and radiation.

LIGHT is an important timing factor for reproductive rhythms (Farner et al., 1966; Jöchle, 1966; Thibault et al., 1966). In wild mammals and birds sexual rhythms will be activated by light in such a way that birth of the offspring will take place during a favourable climatic period. Through domestication an alteration of the correlation between light and reproductive rhythms has taken place. This correlation is still pronounced in domestic animals with seasonal fertility, e.g. sheep, goats, and horses. The fertile period in sheep is initiated by increasing daylight length.

There is a notable variation in the duration of the fertile season within species.

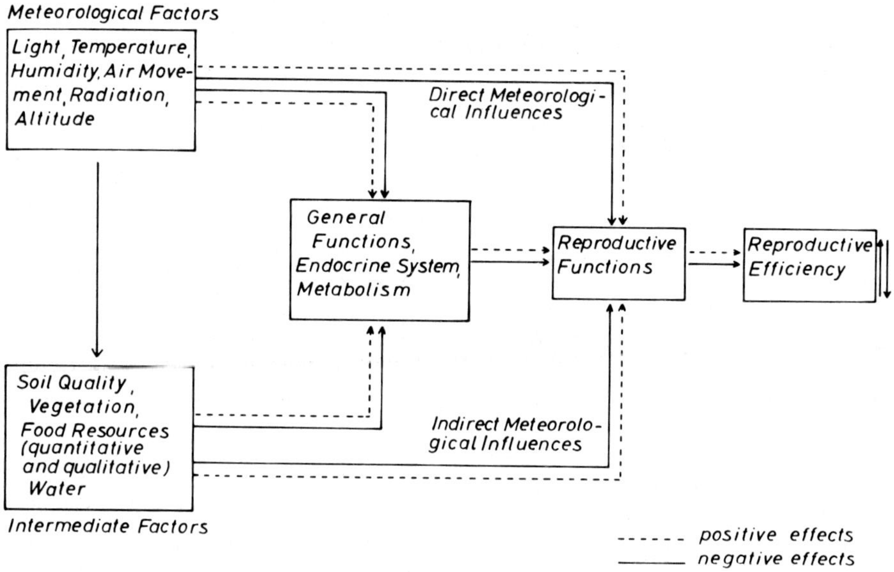

Fig. 1. Meteorological influences on reproductive efficiency

Since the length of the fertile period is partly responsible for reproductive efficiency, light is an important criterion for productivity in the respective species in terms of animal production. Species which are polyestrus throughout the year, such as cattle and swine, exhibit only slight seasonal variations in reproductive traits. In birds laying periods and brooding time are controlled by light. This is especially true for the control of the ovulations sequence by the diurnal light/dark change. The existing knowledge concerning the influence of light on reproduction is utilized, in part, for increasing reproductive efficiency in animal production.

TEMPERATURE influence on reproductive processes must be considered in relation with thermoregulation. Temperatures within the range of thermal neutrality exhibit to pronounced influences on reproductive functions. But those temperatures either above or below the critical temperatures put heat stress or cold stress upon the animal. In this respect the reproductive organs seem to be a "locus minoris resistentiae".

Low temperatures, only below the freezing point, exhibit specific influences or reproductive efficiency. In animals not adapted to cold, low temperatures may negatively affect libido and reproductive efficiency. Local influences of deep cold on the testes may cause disturbance of spermatogenesis and fertilizing capacity. Also, embryonic development may undergo negative influences. Low temperatures are highly critical for newborns because the thermoregulatory mechanisms involved are not yet well established (Yeates et al., 1962).

Natural or artificial shelters against cold diminish the relevance of low ambient temperatures on reproductive efficiency.

More important than low temperatures are high ambient temperatures. Knowledge of their influences on reproduction arises both from field experiences and investigations in climatic chambers under controlled conditions. Generally, extremely high temperatures have a negative effect on reproduction in non-adapted animals (Dutt, 1960; Hafez, 1964; Hafez, 1965). The onset of sexual maturity is often delayed in both sexes under the influence of heat. However, this could partly be channeled via retarded growth. Also, the female sexual cycle can be prolonged or even totally suppressed by heat stress, this is especially known in cattle (Stott et al., 1962).

If copulation or artificial insemination takes place, a reduction in fertilizing rate could occur as well as endangering embryonic development (Dutt, 1964; Smith et al., 1966; Tompkins et al., 1967). Direct temperature effects will be established both through the maternal metabolism and through a direct influence of heat on the first development stages of the embryo (Alliston et al., 1965; Ulberg, 1967). Embryos which overcome this critical period may experience delayed fetal growth due to high ambient temperatures, perhaps resulting in low birth weights (Shelton, 1964; Yeates, 1958). Thermoregulatory problems deriving from lactation, if prevailing synchronously with cyclic activity and conception, may cause additonal negative effects on reproduction.

Very high temperatures will result in decreased sex drive in male mammals. The well known negative effect on spermatogenesis must be interpreted as a local influence of high ambient temperatures on the gonadal tissue and on metabolism (Djanuar, 1965; Waites, 1968). Extremely high ambient temperatures may cause complete cessation of spermatogenesis. Most mammals, however, have local regulatory mechanisms to keep the temperature inside the gonads at an optimal level, that is 3-5°C below the body temperature. In artificial insemination the semen may be exposed to meteorological influences. Techniques utilized in AI guarantee a minimization of effects but in certain cases the success of AI can be reduced. If semen is exposed to high temperatures in the genital tract of females suffering from heat stress, it may retain its fertilizing capacity but cause embryonic mortality (Ulberg, 1969).

In poultry high ambient temperatures result in decreased egg production and egg quality. The hatching ability of eggs will not only be negatively influenced by high temperatures prevailing during semen deposition and fertilization but also by temperatures above the optimal range during the time of storage of eggs until onset of hatching (Hafez, 1968).

Factors other than light, temperature, and humidity are known to influence reproductive processes. These factors include altitude, general and local weather conditions, etc. However, these are mainly of local or temporal importance.

3. INDIRECT METEOROLOGICAL INFLUENCES

Meteorological conditions determine, under natural conditions, the environment for animal populations. Soil quality and vegetation are especially significant here. Concerning natural conditions, the consequences of larger dry periods must be considered. These cause food shortage and decreased food quality. Soil conditions, which are partly due to meteorological factors, influence quantity and quality of foodstuffs of plant origin.

Indirect meteorological influences on reproduction are mediated by the well known relationship between nutrition and fertility. Disturbed reproduction results from starvation and from an insufficient supply of protein, energy, minerals, trace elements, and vitamins as far as they are essential to, or important for, reproductive functions. The supply of drinking water is also of importance.

If animal reproduction is conducted under natural, that is, range conditions, then indirect meteorological influences, as far as reproductive efficiency is concerned, must always be considered.

4. PHYSIOLOGICAL PRINCIPLES OF METEOROLOGICAL INFLUENCES ON REPRODUCTION

Physiological functions are controlled by an integrated system which premarily includes the central nervous system, the pituitary gland, and the gonads. Meteorological influences on reproductive functions are mediated mainly by this control system. Receptors for light, temperature, etc. receive the environmental stimuli which, by neural pathways, are led to the respective central nervous structures. The influences on reproduction are then channeled by neurohormones which are responsible for the gonadotropic activity of the pituitary gland. The gonadotropic hormones influence the functions of the gonads and their target organs.

Meteorological influences may also be mediated through extragenital endocrine systems which control reproduction via the general metabolism. Since the autonomic nervous system not only reacts to meteorological situations but also plays a part in reproductive control, it is also of considerable importance in this context.

Some of the temperature effects are acting locally in the genital organs themselves. The temperature inside the female genital tract will be affected by heat stress conditions of the animals. Male mammals have special thermoregulatory mechanisms to prevent negative local effects of heat stress, such as extracorporal location of gonads, anatomy and function of blood vessels, sweat glands in the scrotal skin, cremaster muscle etc.

Indirect meteorological influences are mediated via general and special genital metabolic functions.

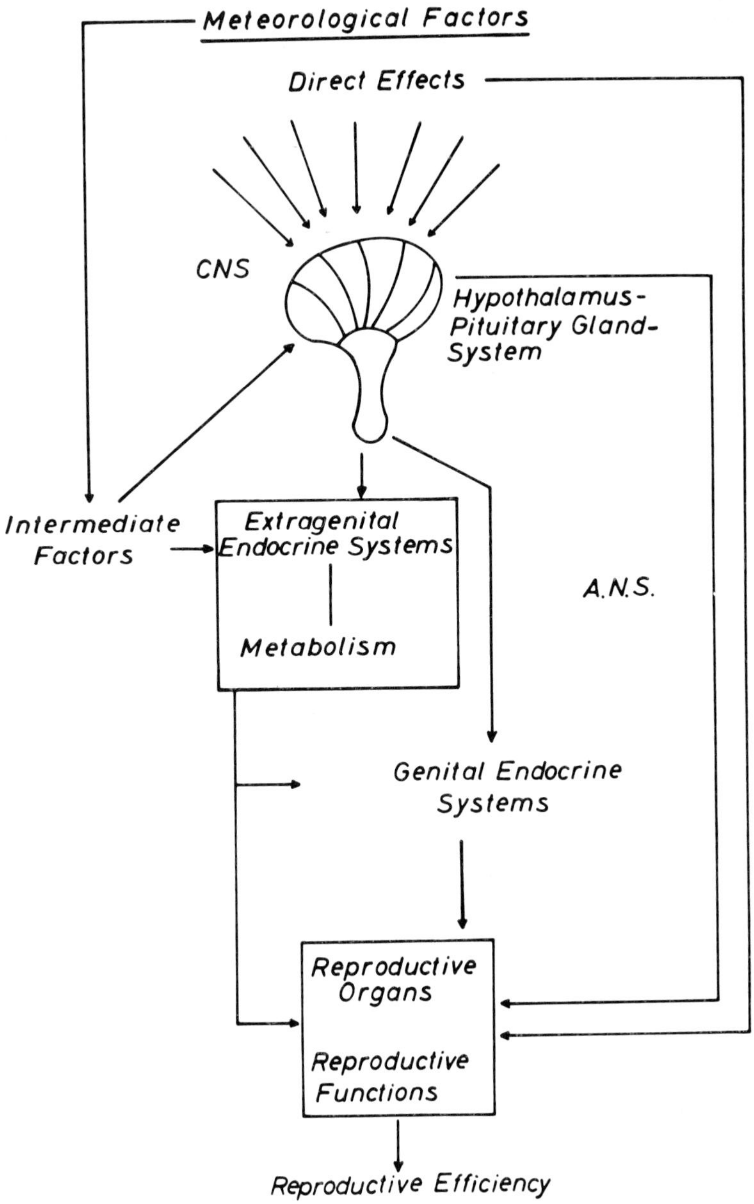

Fig. 2. Physiological principles of meteorological influences on reproduction

In summary, it can be stated that meteorological factors influence reproductive functions in animals, either directly or indirectly. Domestic animals may adapt to extreme climatic conditions and, thus, reproduce normally.

Intensification of animal production requires the use of highly productive animals under unfavourable meteorological conditions. Economy in these cases will often be reduced due to the lack of adaptability. This is especially true of reproductive efficiency. These problems may be partially overcome by breeding measurements, by providing favourable microclimatic conditions for animals and by supplementing the animals food with those nutrients which are lacking in natural resources.

REFERENCES: see pp. 145–146.

Meteorological Effects on Animal Disease

by

W. Ferguson and D. Branagan

1. INTRODUCTION

The manner in which meteorological factors may impair normal function of the animal body is widely variable. Some directly injurious effects are readily discernible. These include such pathological entities as sunburn or frostbite, and such disruption of normal physiological function as marked alteration in level of food intake or failure of normal sweating responses. Where homeostasis becomes disturbed by altered endocrine function, effects such as changes in metabolic rate or reproductive efficiency may take longer to become manifest. When the effect is reduction in resistance to infection it may only be evidenced when the animal's defence mechanisms are challenged by appropriate pathogens.

The overall incidence and spectrum of animal disease is subject to as much indirect as direct influence by meteorological factors. Thus the climatic patterns which govern duration and nature of vegetative growth, and structure, stability and content of soils, will determine the nutritional status of animals which are dependent on natural vegetation for their food supplies. Within a given area, the geographical distribution of essential dietary components such as e.g. drinking water may result in periodic high concentration of animal populations. Such foci of high population densities, by causing increased inter-animal contact frequency, provide conditions highly conducive to spread of infectious disease.

The influence of meteorological factors on the incidence, distribution, and severity of infectious disease should always be viewed in relation to the foregoing background. It is of course governed largely by the nature of the pathogens involved, tending to be least in those diseases which are spread directly from animal to animal and greatest when the pathogen, or its specific vector, is a metazoon which undergoes lengthy exposure to climatic variables.

2. NON-INFECTIOUS DISORDERS

DIRECT PATHOLOGICAL EFFECTS

White varieties of domestic pigs are prone to sunburn in tropical situations when protection against direct solar radiation is inadequate. Frostbite of the comb and wattles of domestic poultry, once common in northern latitudes, has diminished in incidence with the adoption of intensive husbandry techniques. These conditions probably represent the simplest direct relationship between climatic factors and animal health.

Congestive right heart failure with pulmonary hypertension and subcutaneous oedema in cattle at high altitudes is ascribed to low oxygen tension (Alexander et al., 1960). The condition is seen in all breeds, ages and sexes and in both indigenous and imported breeds. The subcutaneous oedema first becomes apparent in the ventral pectoral region, giving rise to the name "brisket disease". Between 0.5 and 2% of cattle residing above 1830 metres become afflicted and a positive relationship exists between the percentage incidence and altitude.

Although resulting from a direct effect of climate, a form of eye cancer affecting cattle exposed for long periods to unmodified tropical or sub-tropical solar radiation differs from the forementioned disorders in that a more clearly defined genetic predisposition is involved in its epidemiology (Freeden et al., 1965). Hereford cattle are more prone than any other breed to this condition which takes the form of a squamous cell carcinoma and is associated with absence of red pigment around the lower eyelid. Anderson and Skinner (1961) have shown that the age adjusted incidence of "cancer eye" in Herefords increased significantly with increase in duration of exposure to sunshine, with decrease in average latitude and with increase in average altitude. The carcinogenic component of sunlight appears to be confined to wavelengths ranging from 2,800 to 3,200 ängstroms in the ultraviolet spectrum.

Harrison and colleagues have studied light induced eye abnormalities in domestic fowl (Harrison et al., 1969). Their results showed that the buphthalmos resulting from exposure to continuous light had a systematic causation.

One of the most dramatic direct effects of climatic stressors on animals is seen in acute heat stress. When the resultant hyperthemia is marked and occurs within a relatively short space of time, the condition is usually accompanied by panting, excessive salivation, restlessness or prostration and is commonly referred to as heatstroke. It may be seen in almost any species of domestic animal and in tropical wild species adapted to forest shade conditions and forcibly exposed to high levels of solar radiation. Animals in transit which are overcrowded, deprived of adequate shade or air movement or, in the case of pigs and buffalo, denied access to water for wallowing for long periods, are particularly vulnerable.

When the heat stress is acute the symptoms are easily recognisable and the condition is somewhat self-limiting since such cases are likely to terminate fatally or to be culled. In commercial livestock production in tropical countries it is the less dramatic, i.e. low-grade chronic cases, which present the more difficult problem. Symptoms tend to be indeterminate and include varying degrees of reduced activity and anorexia resulting in failure to thrive satisfactorily. Such animals become more vulnerable to gastro-intestinal helminths and other pathogenic organisms which may be incriminated in an presumptive diagnosis whilst the underlying malfunction goes unrecognized. At the other end of the scale animals may be subjected to cold stress and when this occurs oxygen consumption increases. When environmental temperature is below the animal's lower critical temperature the rise in metabolic rate associated with maintenance of normal body temperature involves increased energy expenditure. If the food intake does not provide the additional energy requirement for this, the animal utilises body reserves and loses condition.

In tropical and subtropical latitudes poor reproductive performance of domestic animals constitutes an important economic problem for man. Although this is often attributable to low nutritional status, there is growing evidence that high environmental temperature exercises direct depressant effect on sperm production, ovulation and fertility in domestic ruminants and birds and increases the incidence of early foetal death in the former. Since Selye (1936, 1950) focused attention of the influence of environmental stressors on adrenal function, the implications have been extensively studied in laboratory animals and, less so, in domestic animals. Of considerable practical significance for cattle producers in the tropics is the evidence recently produced by Stott (1969) that adrenal cortical secretion of progesterone, in thermally stressed cows, results in markedly elevated blood levels of this hormone and that this effect can be prolonged for some time following cessation of the thermal stress. This could prove to be an important causal factor of anoestrus in cattle in hot environments.

Non-sweating (syn. Anhidrosis) of horses occurs in the humid tropics. The physiology of sweating in horses has been studied by Evans and co-workers who showed that sweating in this animal is stimulated by adrenalin and not by the sympathetic nervous system (Evans et al., 1957). It is postulated that sustained exposure to high environmental temperature and relative humidity gives rise, in the horse, to sustained release of adrenalin to which the sweat glands become insensitive following prolonged overstimulation. The condition constitutes severe impairement of a horse's capacity for heat dissipation thus rendering it more susceptible to acute hyperthermia if subjected to violent exercise. It has not been reported from the drier tropics where climate is conducive to efficient evaporative cooling.

NUTRITIONAL EFFECTS

Animals, particularly herbivores, are largely dependent on the quantity and quality of vegetative growth for their nutritional well-being. Where rainfall or warmth are insufficient to support vegetative growth for lengthy periods of the year they commonly lose condition rapidly during the quiescent periods. Such losses are often severe in countries where education, land tenure systems and economic considerations have not generated adoption of fodder conservation and supplementation on a significant scale.

At the other extreme i.e. where conditions of temperature and soil moisture are adequate to maintain vegetative growth throughout the year, the quantitative yield of animal fodder may be constantly assured yet its quality may become inadequate to maintain good health. Available nitrogen in the soil is often very limited and as plants mature their aerial portions require an increasing measure of rigidity to maintain the necessary upright stance in their environment, i.e. the rate of protein dilution and lignification in the plant tends to be directly proportional to the rapidity of growth. Thus in the humid tropics animals are often unable to ingest sufficient dry matter to secure the energy and protein requirements for maintenance of good health.

Leaching of minerals from topsoil is more likely to occur under conditions of high rainfall and humidity. In drier climates a higher proportion of surface water is shed by evaporation which leaves a higher concentration of salts near to the soil surface. In addition, plants adapted to the drier areas are generally endowed with deep root systems capable of bringing salts from the deeper soil layers to the surface. In practical terms this natural relative advantage of the drier climates is offset by the fact that the high yeild potential of the warm high rainfall areas offers better prospect of economic return from artificial fertiliser application.

In temperate climatic regions fodder conservation and supplementation, and improvement by fertiliser application are now well advanced yet relationship of climate to mineral deficiency diseases is still apparent, albeit complex, as evidenced by e.g. the seasonal peaks in incidence of hypomagnesaemia ("grass staggers"). The clinical condition in cattle and sheep is typically seen when these animals are put out to graze lush young pastures in spring or autumn. Magnesium is concentrated in the seeds of plants and more especially in the seed coatings. Young, relatively shallow-rooted pasture plants have a lower magnesium content than mature deep rooted grasses or legumes. However this is not the principal factor in the aetiology of hypomagnesaemia since the fall in plasma magnesium is too rapid for this to be so and since the condition can occur in animals fed grass with normal magnesium content. Absorption of magnesium from the alimentary canal appears to be impaired when young lush grass is fed. It has been shown that less magnesium may be absorbed from the latter than from hay fed in amounts which supplied the same daily quantity of magnesium (Care, 1967). Relationship between calcium and magnesium absorption rates appear to be involved in the aetiology and research attention is currently focused on the possible existence of a natural agent present in young spring grass which interferes with absorption of these minerals from the alimentary tract. Hay and silage conserved early in the season i.e. derived from young grass, may be relatively deficient in magnesium and calcium and animals fed on this during the winter months are likely to be particularly susceptible when turned on to new grass in spring.

The whole-plant concentration of phosphorus and potassium declines markedly with advancing maturity. In the case of the micronutrient elements valid generalisation cannot be made. Experimental results have been so variable and so frequently conflicting as to lead at least one authority (Underwood, 1962) to conclude that variations in total trace element concentrations in mixed herbage due to the effects of season are not of themselves likely to be of large and widespread significance to grazing stock. On the other hand Ollerenshaw and Smith (1969) have drawn attention to the likelihood in Great Britain of a relationship between incidence of "swayback" and the number of days with snow covering the ground. Swayback, a nervous disorder of new-born and young lambs, is associated with copper deficiency in the pregnant ewe although there is evidence that a low copper status alone does not necessarily cause the condition.

The metabolic disorders of sheep which may be precipitated by change in weather include pregnancy toxaemia. This highly fatal disease of late pregnancy was in earlier years referred to by the appropriate name "snow blindness" because of the high incidence in heavily pregnant ewes when their food supply was rendered inaccessible by a heavy fall of snow. The disease is essentially a carbohydrate depletion induced by the high demands of pregnancy, particularly in ewes carrying twin lambs, and exacerbated by any interruption in food supplies or exposure to sudden cold weather.

Avitaminosis A occurs widely in domestic ruminants in climates characterised by a long dry season and associated paucity of green food. The condition accounts for reproductive disorders including abortion and, in more general terms, for impairment of resistance of epithelial membranes to infection.

3. INFECTIOUS DISORDERS

HOST RESISTANCE TO INFECTION

The extent to which meteorological factors may determine the severity of certain infections is now widely acknowledged.

Andrews (1964) in discussing the epidemiology of respiratory virus infections in humans suggests that a state of equilibrium may be attained between host and infective organism wherein the latter does not become significantly invasive or pathogenic until the balance is upset by an effective stressor associated as a rule with some meteorological change to which the host body does not adjust quickly enough. Similar host-parasite relationships are known to occur within animal populations. Schope, quoted by Andrews, was able to activate latent swine influenza by wetting pigs with a hose, Bierer (1962) induced clinical *Escherichia coli* infections in chicks by chilling the latter artificially, while Scholes and Hutt (1942) demonstrated a close relationship between body temperature and resistance to *Salmonella pullorum* in the domestic fowl.

The influence of temperature on host resistance to infection had been demonstrated almost a century ago by Pasteur and co-workers (Pasteur et al., 1878) who succeeded in infecting the domestic fowl with *Bacillus anthracis* after inducing hypothermia in their subject. More recently an attenuated strain of *Ectromelia* virus has been shown to be at least a hundred times more lethal for mice kept at 2°C than it was for mice kept at 20°C (Roberts, 1964). The author of the latter report suggested that the mice maintained in the low temperature environment depend on an increased metabolic rate which renders them susceptible to liver damage. These exists some evidence that exposure to high environmental temperature may also be associated with lowering of resistance to certain infections. Thus Sinha and associates (Sinha et al., 1957) found that seven week old chickens succumbed more readily to Newcastle Disease virus when exposed to high environmental temperature than when in a cold environment.

In those experiments involving induction of hypothermia or hyperthermia in the host it is not clearly established whether the alteration in host body temperature per se served to provide a more favourable environment for the pathogen employed or whether it served to inhibit the host's defense mechanisms. Adrenal activity is known to be altered by a wide variety of stressors (Selye, 1950) including heat (Stott 1969) and cold (Robertshaw et al., 1969) so that involvement of host defense responses seems likely.

Environmental temperature may influence the primary defenses of the respiratory tract in a physical manner. Low environmental temperature tends to inhibit ciliary movement whilst it seems likely that high environmental temperature, when it results in panting involving mouth breathing, may, by omission of the nasal barriers, cause admission of larger particle size and heavier infective loads to the bronchial and perhaps to the alveolar level of the respiratory tract.

The incidence of respiratory diseases in animals appears to be modified by atmospheric humidity. To gain access to host tissue, infective material must penetrate the mucous membranes lining the respiratory tract. These surfaces are partly protected by their mucous secretions which contain non-specific anit-microbial substances such as lysozymes and these defenses are reinforced by the ciliary barrier of the lining epithelium in the upper respiratory tract. When the inspired air is of very low relative humidity some drying of the upper respiratory mucosa, arrest of secretory flow, and inhibition of ciliary activity may occur, with consequent increased risk of microbial infiltration and infection.

The physiological significance of low atmospheric humidity for rats has been described by Dalham (1956) and there exists evidence of similar influence on the incidence of clinical respiratory infections in domestic animals. Ekmeier and Agde (1965) confirmed to their satisfaction that fall in relative humidity triggered off clinical outbreaks of respiratory disease in housed pigs. Sainsbury (1968) has suggested that the relative humidity of poultry houses should be kept as high as 60 per cent to provide the best environment to enable birds to resist the effects or respiratory infections, and Tripp (1965) states that in the view of many experienced poultrymen an adequate level of humidity is required to minimize risk of avian infectious bronchitis. These views are supported by the high incidence of respiratory infections in poultry in countries with a hot-dry climate, although in the latter a high atmospheric dust content is often a contributory factor.

PATHOGENS AND THEIR VECTORS

The influence of gross climatic factors upon the distribution of disease tends to be least when the pathogen involved is a directly transmissible virus, rickettsia or bacterium, and tends to be most marked when the pathogen or the specific vector is a metazoon whose essential life-cycle entails a considerable period of exposure to those gross climatic factors. Hence, the present general distribution of many of the major viral, rickettsial and bacterial diseases has arisen from differing emphases in veterinary control rather than from different climatic influences. Outstanding examples of this broad situation are provided by foot and mouth disease (FMD), rinderpest and contagious bovine pleuro-pneumonia (CBPP).

Though it exists endemically in both temperate and tropical latitudes, FMD is decidedly more prevalent in the latter areas, and it can be safely assumed, for instance, that there is no period of any year in which the cattle populations of East Africa are entirely free from the disease. This difference is an artifact due to the greater pressures of veterinary control which prevail in most temperate countries, an effect which has been even more pronounced in the cases of rinderpest and CBPP. Both of the latter diseases were once endemic in Europe and are now largely confined to residual reservoirs in those tropical countries where effective measures could not be imposed.

The fact that conventional control measures were successful in eliminating rinderpest and CBPP in Europe and did not succeed in the case of FMD is indicative of the factors which influence the precise distribution of disease within a general pattern. Hence, while such measures as the prohibition of stock movement and the imposition of orthodox sanitary procedures are frequently successful in controlling the spread of CBPP and rinderpest, recent studies on the airborne dissemination of FMD have offered a convincing explanation of the failure of such methods to prevent the spread of this highly infective virus.

The FMD epidemics in England between 1952 and 1968 are described by Hurst (1968) and by Smith et al., (1969), and the accumulated evidence shows that dissemination from a primary focus usually occurs downwind (upwind in only 8.6% of cases), and is particularly liable to occur at night, presumably since the airborne particle is not then exposed to ultra-violet light. Periods of highest infection rating occurred during heavy rain, and, where peak incidences occurred during dry weather, the associated epidemics were on a much smaller scale. Distances between primary and secondary outbreaks were less than 20 km in 91% of cases. All this accords well with physical studies on the airborne dispersal of small (i.e. $7 - 10\mu$) particles and the deposition of these particles during rain.

Tromp (1968) and Hyslop (1969) provide excellent sources of references to this and allied phenomena, and the latter described work which demonstrated that the longevity of rinderpest virus is greater at low humidities. Hence, the association of increased rinderpest-spread patterns during drought in East Africa (Branagan et al., 1965) may not be a pure function of large stock concentrations around the few remaining waterholes, but may be a complex function which includes a heightened survival rate of the virus outside the host.

An interesting corroboration of Hemmes' (as quoted by Tromp, 1968) criteria for grouping micro-organisms according to their environmental requirements is given by the distribution of anthrax. Endemic in most tropical and sub-tropical countries, this disease tends to occur sporadically (and only occasionally in epidemic form) in temperate countries of high latitude. While the above-mentioned influences of differing standards of control and sanitation have an undoubted and powerful effect within this phenomenon. Minett (1952) has drawn attention to a critical minimum temperature (around 21°C) for the sporulation, and hence the prolonged ex-host survival, of *Bacillus anthracis*. For this reason, the sporadic occurrence of anthrax in northern European countries is most frequently associated with the winter feeding of supplementary materials imported from tropical countries. It is noteworthy that one of the rare epidemics in Norway and Sweden occurred during the abnormally hot summer of 1937.

Evidence is continually accumulating upon the manner in which annual climatic variations may influence the seasonal incidence of non-metazoon, directly-transmissible pathogens within a particular locality – i.e., the influence being on the incidence rather than upon the general distribution. A rise in coccidiosis incidence in abnormally wet years has been traditionally associated with a lowered mortality among oocysts shed upon saturated pastures and with a greater degree of contamination of water sources by oocysts preserved in the mud on the feet of livestock (Davies et al., 1963). In Britain, Hugh-Jones and Austwick (1967) have recognised a positive correlation between the number of days of rainfall in June and the incidence of bovine mycotic abortion (*Aspergillus* spp) during the subsequent 12 months, and have contrasted this with the absence of such a correlation involving the September rainfall. The majority of such studies can, however, be regarded as being in a preliminary and exploratory stage, and much further work requires to be done at microclimatic levels of precision before the implications of the patterns can be understood. This structure applies equally to work on pathogenically significant metazoa.

There are many instances where gross climatic factors determine the distribution of pathogenically significant metazoa. Obvious examples occur, of course, among those diseases whose distribution is strictly determined by the existence of an arthropod vector e.g. malaria by mosquitoes, trypanosomiasis by tsetse fly, and various protozoa by ixodid ticks.

Among helminths, a definite climatic limitation in distribution is exemplified by *Gaigeria pachyscelis* and *Mecistocirrus digitatus.* Both of these nematodes, (the first affecting the duodenum of sheep and goats in India and Africa and the second the abomasum of cattle, buffalo and sheep in South America, India and the Far East) are strictly confined to tropical latitudes, and it is thus assumed that their requirements for off-the-host survival are not met outside these geographical limits. Conversely, *Ostertagia circumcincta,* a nematode affecting the abomasum of sheep and goats, is most prevalent in the higher latitudes, and its incidence declines in both hemispheres as the equator is approached, e.g., it is common in South Africa, but becomes progressively less so as one moves north towards the Central African states. It is postulated that these distributions are largely governed by two factors which have a dominant influence on the off-the-host stages of the organism. The first of these is the influence of temperature upon the rate of development; and the second is the influence of saturation deficiency (i.e. the combined effects of temperature and humidity) upon survival.

Silverman et al (1959) showed that the development of the eggs of *Haemonchus contortus* (a nematode causing parasitic gastro-enteritis of sheep and goats) was almost twice as fast as that of *O.circumcincta* eggs at 7°, 11° and 14°C, and hence the increasing incidence of the former with increasing mean minimum temperatures is explained by a greater production of infective larvae per unit time in warmer climates. With the rise in mean minimum temperatures, however, there is an accompanying overall rise in the lethal, desiccating effects of raised saturation deficiencies, and the disappearance of *O.circumcincta* at the lower latitudes may well be decided by this particular influence. The interaction of these two factors is often demonstrated by events following unusual climatic conditions within a specific locality – as, for instance, by the so-called "haemonchosis storms" in East Africa.

Though infestations of *H. contortus* are a common clinical entity in sheep and goats in the highlands of East Africa during the warm, rainy season, the problem only assumes serious epidemic proportions in abnormally wet years. If the reason for this, as is commonly supposed, is that the very prolific and rapid development of infective larvae brought about by the high temperatures is countered during normal years by a high mortality of larvae due to a raised saturation deficiency created by these high temperatures, then the severe losses encountered during abnormally wet years is explained by the partial removal of the limitations of saturation deficiencies on the survival of the larvae. It is highly probable that these processes are influential in deciding the distribution and incidence of most nematodes, certainly with such species as *Trichostrongylus colubriformis, Bunostomum phlebotomum* and *Neoascaris vitulorum* which show a markedly greater degree of activity in warmer climates.

An association between abnormally wet summers and an increased incidence of fascioliasis (due to *Fasciola hepatica*) in Britain during the following autumns and winters has long been recognised. This association is a function of, first, the necessary temperature level (10°C and above) for development outside the final host; and, second, the wet pastures which provide improved conditions for the proliferation and survival of the intermediate host (the snail *Lymnaea truncatula).* It has thus been possible for Ollerenshaw (1966) to use summer meteorological data as a basis for forecasting the probable incidence of fascioliasis during the following autumn and winter. A similar empirical system has been devised (Ollerenshaw et al., 1966) for forecasting the incidence of *Nematodirus battus*, cold winters and late springs being associated with a high incidence of nematodiriasis, whereas mild winters and early springs result in a peak incidence of incidence of infective larvae occurring on pastures before lambs have begun to consume large quantities of grass.

Specific habitats of intermediate hosts, such as those required by snails in the life cycles of *Fasciola* spp, are frequently dependent upon general topgraphical features, and thus the local epidemiology of disease may be crucially influenced by climatic events which take place at some distance from the focus of disease. Excellent studies on the epidemiology of fascioliasis in East Africa by Coyle (1961) and Hammond (1965) have stressed the significance of spring-fed pools in the proliferation and survival of the snail *Lymnaea natalensis* (the intermediate host of *Fasciola gigantica*). Obviously, rainfall in mountain areas at some distance from these pools may well decide both their survival and that of their dependent snail populations. Further, as indicated by Hammond (1965), the outflow from these pools can be responsible for the dispersal of infected snails over the lower ground. Such dispersal could, for instance, bring about an annual reinfection of lower ground, or a permanent establishment of infection in the event of the water-supply being used in an irrigation scheme. Conversely, sudden increases in water flow often destroy snail habitats on lower ground, either by scouring these out or by depositing heavy layers of mud and sand.

The probality that these two factors of temperature and saturation deficiency are responsible for confining the trypanosome-bearing genus *Glossina* to a belt running from east to west across the continent of Africa is indicated in the classic and comprehensive work by Buxton (1955). Extension northwards from this belt is prevented by excessive aridity, and, southwards, it appears as though declining minimum temperatures at first retard and finally disallow succesful development. Within these limits, the distribution of the genus is broadly restricted to areas with a mean annual rainfall of not less than 500 mm. Thereafter, the degree of infestation by specific members of the genus is decided – again, on broad terms – by the local climate and the type of vegetation which this climate supports.

Most of the *fusca* group of *Glossina* are regarded as "forest species" and make up typical infestations of rain forest and humid thicket in which seasonal variations in activity are not strikingly marked (though activity may be restricted or absent during the mid-day periods of the hot season – Page, 1959; Jordan, 1962, 1963). On the other hand, infestations of the drier parts of the general distribution, usually by the *morsitans* group, tend to be confined to the isolated patches of tree or shrub vegetation typical of such areas and to show marked seasonal variations in activity, i.e. peaks of activity during the warm, wet season declining to low or non-existent levels during the cool, dry season. These dry areas also show a characteristic ebb and flow of local distribution over periods of years. Thus, after a number of years of low rainfall, the more isolated tsetse colonies tend to die out, so that relict populations are left only in those larger patches of cover which often afford a greater degree of host opportunity, e.g. trees around a water hole. Given a series of years with high rainfall, however, these relict populations tend to expand and to recolonise the more isolated patches of cover.

At the peripheries of *Glossina* distributions, there is often a very specialised dependence upon the existence of a particular microclimate as, for instance on that which is provided by a particular species of shrub. The maintenance of a *Glossina* spp in these circumstances is inclined to be precarious, and many examples exist where tsetse have been cleared from large tracts of country by, for instance, the simple removal of a particular species of shrub. Conversely, many attempts at clearance along these lines have not been successful, and in such cases it has been obious that the requisite microclimates were afforded by unsuspected alternative circumstances. Besides the known influence of edaphic factors on *Glossina* puparia (Nash, 1939), the critical conditions for survival of tsetse populations are met at microclimatic levels which bear a variously elastic relationship with gross climate.

Arthur (1962) states that "ticks surpass all other arthropods in the number and variety of diseases which they transmit to man and his domestic stock", and an additional appreciation of the scope of these relationships is gained by realising that almost every environment capable of supporting domestic livestock is also capable of supporting a population of at least one species of Ixodoidea. The great range of adaptation to climatically influenced factors is, however, principally exhibited by the hard ticks (Ixodoidae), for the soft ticks (Argasidae) are usually xerophilic and tend to be confined to the same sort of habitat and life cycle wherever they occur, i.e. they are essentially lairage parasites, inhabiting crevices in the walls and floors of human and animal resting places and issuing from these retreats to make frequent, swiftly-completed blood meals. Often acting as significant vectors of disease (e.g. as in the *Argas persicus* – fowl Spirochaetosis relationship), the influence of climatic factors upon their incidence is relatively uncomplicated within their overall geographical distribution and might be said to be confined to a choice of dry retreats in very humid conditions, or a choice of moist retreats in excessive aridity.

Although the essential biology is similar in all ixodid ticks, the extent and character of their life-cycles is largely determined by the length of their developmental and waiting periods. Since the duration of developmental periods is directly influenced by temperature, and since survival during both developmental and waiting periods is influenced by levels of saturation deficiency, the effect of climate will decide, first, whether or not a particular species of tick (and hence a particular tick-pathogen relationship) can persist in an area; and, second, the pattern of overall and seasonal incidence of the tick and its dependent disease.

In Europe and Russia, the genus *Ixodes* is primary significance in the transmission of disease, *I.ricinus* for example being a principal vector of babesiosis, of tickborne fever of sheep, and of a number of virus infections such as louping ill and spring and summer encephalitis. Possibly more detailed attention has been given to the ecology of *I.ricinus* than to that of any other ixodid species, and the work so produced by British and Russian authors forms a sizeable part of the classic literature on ticks (for example, MacLeod, 1932, 1939; Milne, 1946, 1947, 1948; Kheisin et al., 1956; and many others). The marked vulnerability of the species to low humidity confines it in Britain to vegetation typified by the presence of a moisture-retaining substrate (the so-called "basal mat"), and the protracted life-cycle, with one year elapsing between the engorgement of one instar and the emergence of the next, is regulated by diapause mechanisms which delay development so that activity at unfavourable seasons is avoided. Though still not fully understood, these mechanisms involve an interaction between declining or increasing day-lengths and declining or increasing temperatures. The result of these climatically-regulated controls is that, in most parts of Britain, the seasonal activity of *I.ricinus* follows a bimodal pattern with peak incidences occurring in spring and autumn, while, in Russia, there is a unimodel summer peak. The epidemiological patterns of diseases borne by *I.ricinus* naturally conform to these seasonal distributions of tick activity.

Approximations to this "basal mat" which is typical of *I.ricinus* habitats in Britain are significant in the maintenance of other disease-bearing tick populations. Well-described examples occur in South Africa (Stampa et al., 1958), where *Ixodes rubicundus* prevails in the coarse *Danthonia-Aristida* herbage on the southern aspects of hills; and in Canada, where the commonest habitat of *Dermacentor andersoni* is among the decomposing litter at the base of pastures dominated by *Agropyron* spp (Gregson, 1951). Both of these tick species are associated with tick (or Karroo) paralysis, and, in each case, the disease is associtated with only certain parts of the ticks' distributions. The precise mechanism in these particular tick-associated paralyses is not known, but two possibilities may be tentatively considered in view of their distributional inconsistencies. The first possibility is that the disease-inducing character of the ticks arises from a particular previous experience of hosts with a specialised distribution. The second possibility is that microclimatic variations in different plant communities may influence the developmental metabolism of ticks to a degree which determines whether or not the emerged instar can induce paralysis, possibly by the biochemical nature of their attachment and feeding secretions.

As far as is known at present, the life-cycles of tropical and sub-tropical ticks are usually uncomplicated by diapause mechanisms. Rates of development bear simple relationships to temperatures so that, by this alone, peak incidences of ticks tend to occur as their emergence is accelerated during the warmer periods of the year. As saturation deficiencies assume greater significance in the drier sections of tick distribution, so the periods of maximum activity become more strictly confined to the wet season, for it is only at this time of the year that prevailing saturation deficiencies will allow the survival of the more vulnerable stages of the tick (i.e. the eggs, and emerged larvae and nymphs). A typical pattern of a tropical tick's life cycle and distribution in relation to climate is given by *Rhipicephalus appendiculatus.*

The relationship between *R.appendiculatus* and the causal organism of East Coast fever, the protozoon *Theileria parva*, is probably the most intractable animal health problem in Africa, south of the Sahara, and it is accepted by most authorities that the continuing presence of the disease virtually confirms the presence of its specific vector. Endemic East Coast fever is thus confined within the normal distribution of *R.appendiculatus* which, in East African, is limited to areas with a mean annual rainfall of at least 630 mm. There is a considerable variation in climate and dependent vegetation within this distribution, and accompanying variation occur in the life cycles and incidence of localised *R.appendiculatus* populations with resultant influences upon the local epidemiology of East Coast fever.

The cooler, more humid extremes of the distribution occur in the East African highlands, where the climatic conditions can be crudely indicated by a mean annual rainfall of 1,000 mm, and by mean, mean maximum and mean minimum temperatures of 15°, 20° and 10°C. Grass cover tends to be a dense sward. In such circumstances, the temparatures allow only one life cycle to be completed each year. Humidities are always sufficient to allow the survival of eggs and larvae, and hence a life cycle can be completed in such an area no matter at what time of the year it is commenced. Though emergence is accelerated and reaches a peak during the warm season, neither this rise nor the corresponding decline in the cool season are as marked as in other parts of the distribution. Hence, apart from a variation in degree, the possibility of *R.appendiculatus* acquiring and transmitting *T.parva* continues throughout the year.

At the other extreme the crude climatic conditions would be indicated by a mean annual rainfall of 650 mm, and mean, mean maximum and mean minimum temperatures of 18°, 24° and 11°C. The grass cover is very much less dense than the previous example. The temperature in such an area allows for two life cycles to be completed each year, and the fact that this seldom occurs is due to the limiting effect of humidity on survival. As a rule, the decisive factor in the survival of a *R.appendiculatus* population is such an area is whether the degree and duration of favourable humidities during the wet season are sufficient to allow the hatching of eggs (conditions for the hatching of eggs are never favourable in such places during the dry season). *R.appendiculatus* populations which survive in such areas usually show a marked seasonal activity, with adults appearing at the start of the wet season followed closely by larvae and then nymphs. Activity tends to be minimal during the dry season, with the tick population passing this period as engorged nymphs whose emergence as adults is delayed by the seasonal coolness until temperatures start to rise at the commencement of the wet season. In the relatively thin grass cover typical of such areas (i.e. much less dense than in highland areas), the soil surface upon which eggs are usually deposited dries out quickly so that humidities within the grass cover are thereafter maintained above general atmospheric levels only for as long as grass transpiration continues. As evaporative stress reaches certain critical levels during the heat of the day, so the transpiration of xerophilic and semi-xerophilic grasses is arrested, and humidities within this density of grass cover rapidly achieve equilibrium with those of the general atmosphere. This result will swiftly desiccate *R.appendiculatus* eggs. Though nearly as vulnerable to aridity as eggs, larvae rest on the undersides of leaves within the direct humidity of transpirational microclimates. Thus, they exist in higher humidities than eggs while transpiration continues, and benefit from an immediate restoration of high humidity when transpiration recommences when the period of evaporative stress has passed (Branagan, 1970).

This critical limiting factor in the distribution of *R.appendiculatus-T.parva* relationships offers an explanation for the appearance of epidemic outbreaks of East Coast fever in normally unaffected areas. Unusually high rainfall in such areas allows a sufficient duration of humidities which allow eggs to hatch, and, once this has happened, a *R.appendiculatus* population may well be able to persist until a restoration of normally arid conditions again disallows hatching. Further, the high temperatures common to such areas accelerates development so that, given adequate host opportunity, the life cycle moves rapidly through to the emergence of adults, which is the emerged stage least vulnerable to dessication and which is probably the most effective in the transmission of *T.parva*. This interpretation agrees with the findings made during the extensive and detailed field studies by Yeoman (1966, 1967).

In East Africa, recent studies on potential evaporation (Whoodhead, 1968a, 1968b) have been made in conjunction with work on the reclassification of rangeland (EAAFRO,* to be published).

This revealed that the rate of evaporation in relation to the volume of available moisture was of critical importance in deciding the composition of plant communities, and was often influential in deciding the precise physiological performance of the plants established. Evaporation rather than rainfall is thus seen to be a factor of great ecological significance in the distribution of vectors such as ticks, for the survival of these is largely dependent upon microclimates within a vegetative cover, and these, in turn, can be decided by the physiological performance of the plants concerned.

Classic examples of climate exerting a limiting effect upon an arthropod-borne

* East African Agriculture and Forestry Research Organisation.

pathogen rather than upon the vector occur in various *Anopheles-Plasmodia* relationships where the development of the sexual cycle of the malaria parasite in a mosquito is arrested at temperatures below 16°C and recommenced when this level is exceeded. On one hand, development may be suspended over a period of hibernation and resumed when the mosquitoes regain their activity in the spring. On the other hand, the development of *P.vivax* within *A.maculipennis* becomes indefinitely prolonged at low temperatures, so that the parasite eventually dies in the mosquito without reaching the infective stage (Dogiel, 1964). Thus , situations exist in temperate climates where malaria is not normally transmitted even though suitable *Anopheles* vectors co-exist with malaria carriers, but where transmissions are often effected during abnormally hot summers. A similar situation exists in the *Aedes aegypti*-yellow fever relationship, where the mosquitoes' ability to transmit the virus is suspended at temperatures below 18°C (Pavlovski, 1947).

Lutta and Kheisin (1954) describe an apparant influence of climate upon the efficiency of *Ixodes persulcatus* as a vector of *Babesia bovis* whereby the incidence of clinical babesiosis is more severe in the southern Russian sectors of this relationship's distribution than in the north, and tentatively attribute this difference to development of the protozoon within the tick being inhibited at lower temperatures. It has also been shown that the marked contraction in *Rhipicephalus appendiculatus* development periods which is achieved at higher temperatures is accompanied by a loss of ability to transmit *Theileria parva* (Fotheringham et al., 1937). Hence, there are grounds for believing that climatic factors may impose upon ticks rates of development which are "out of phase" with the rate of development which these factors impose upon protozoa within the ticks, so that infectivity is either reduced or destroyed. A possible explanation is thus offered for those situations sometimes described in field observations where tick vectors and mammal hosts co-exist without any recorded incidence of an associated disease, e.g. *R.appendiculatus*, cattle and an absence of East Coast fever.

Much of the basic information on the survival of pathogenically significant organisms is derived from experiments carried out under constant conditions. As such, it is frequently of limited value when applied to field conditions, and particularly so when applied to metazoa. For example, the survival of ticks is largely determined by water relations, water being lost or regained according to revailing levels of saturation deficiency, and lost or regained more rapidly at higher temperatures. Thus, the diurnal variation profiles of temperature and humidity are of critical importance to the survival of such organisms. Moreover, the variations relevant to the organism concerned will tend to take place at microclimatic levels, often within 2 or 3 mm of a plant surface. This situation presents serious technical difficulties in the recording of precise microclimatic data, and hence it is probably true to say that studies on the effect of climate on a large number of pathogenetically significant organisms will remain at an empirical level until this problem is solved.

DISEASE EXPOSURE RISK

It is well established that wildlife populations may constitute important reservoirs of diseases transmissable to domestic animals. The practical feasibility of utilising marginal rangeland for domestic stock raising is often determined by climatic factors and, in particular, rainfall patterns. Thus in certain semi-arid areas of Africa the composition and density of vegetation, and the distribution of surface water, may be eminently suited to certain forms of wildlife survival yet inadequate for continuous exploitation by man and his domestic herds or flocks. The improvement following good rainfall may attract the latter into such areas temporarily, thus bringing them into closer contact with the wildlife. The preclusion of permanent human settlement favours the survival of the wildlife populations and climatic variation can influence the degree and periodicity of contact between domestic herds and these potential reservoirs of such epidemic diseases as rinderpest, trypanosomiasis and salmonellosis.

Rainfall pattern can be an important determinant of the rate of spread of epidemic disease within domestic animal populations. In tropical and subtropical latitudes, periods of surface water scarcity lead to large scale stock dispersals from dried out watering point vicinities and concentrations around the reduced number of persisting watering points.

This greater contact frequency, coupled with greater water contamination risk, has proved to be a major determinant of rapidity of spread of *Salmonella typhi-murium* infection in sheep in New Zealand (Jebson, 1950) and in Australia (Watts et al., 1952).

Animal compensatory behavioural response to climatic factors may have epidemiological implications. High levels of solar radiation, by stimulating shade-seeking, may bring animals into closer contact with the tsetse fly vectors of trypanosomiasis, and with infective helminth larvae. Similarly, high effective environmental temperature stimulates buffalo to wallow in surface ponds or streams and may thus bring this species into closer contact with the snail intermediate host of liver fluke.

REFERENCES: see pp. 146-148.

Meteorological Effects on Arthropods (Including Insects, Ticks and Mites) as Vectors of Human and Animal Diseases

by

W.O. Haufe

1. INTRODUCTION

Weather and climate have been shown in several previous reviews (Graham, 1956; Haufe, 1963a, b, 1964a; Johnson, 1954; Messenger, 1959: Uvarov, 1931 and Wellington, 1957) to influence the abundance and activity of arthropods. Many species incriminated as vectors of disease depend on critical relationships among environmental conditions to function as effective transmitters of pathogenic organisms among animals and man (Haufe, 1964a). Meteorological factors affect distribution and numbers of disease vectors directly through their influence on rate of development, longevity, fecundity and regulation of behaviour patterns that are essential for survival and propagation. Although feeding of vectors on man and animals is often incidental or accidental, the host-parasite association as well as the availability and accessability of other sources of food may depend indirectly on the meteorological environment. Temperature, moisture, light and wind collectively play major roles in limiting dispersal of arthropods, in modifying patterns of displacement to areas beyond their breeding grounds, and in regulating geographic ranges for endemic and epidemic spread of vector diseases in the human and animal populations.

The earlier work tended to concentrate on analyses of cause and effect between individual meteorological factors such as temperature, humidity and radiation on the one hand, and biological processes such as growth, activity and longevity on the other (Uvarov, 1931). Interpretation of biometeorological relationships in the ecology of insects has been generally referred to selected areas of study such as forest insects (Graham, 1956), bioclimatics (Messenger, 1959), synoptic concepts of insects and climate (Wellington, 1957), migration of aphids (Johnson, 1954) and insect vectors of disease (Haufe, 1963a). More recent biometeorological research has indicated a high degree of interaction among physical factors in the over-all influence of weather and climate on the response of insects (Haufe, 1963b), and conventional statistics have been inadequate so far in the development of methods to predict infestations. Although multiple and partial regression analyses have been useful for evaluating long-term relationships between climate and abundance (Williams, 1958, 1961), they have been relatively inconclusive in schemes to describe dispersal activity and the resultant short-term changes in spatial and geographic distribution of populations.

Biometeorological studies on arthropods during the last decade have shown the need for a new approach emphasizing integrative investigations. The literature contributing to further progress in environmental studies has been reviewed mainly for the period 1963-70 to consolidate the knowledge of significant biometeorological processes pertaining particularly to species of veterinary-medical importance. Some concepts have significant support from isolated investigations to permit tentative integrative interpretations of direct influences on the human environment.

2. HYGROTHERMAL REGULATION OF METABOLISM AND RELATED ACTIVITIES

No relationships established between single meteorological factors and vital processes of development and behaviour are consistent for a broad range of arthropod species. There is recent evidence, however, to indicate that water balance between the internal and external environments is a consistently well-developed feature of physiological structure in many unrelated species. Hygrothermal regulation is the essential predominating mechanism in mosquitoes maintaining favourable water balance between tissues and the atmosphere and facilitating environmental survival (Haufe, 1966b). Activities and longevity depend on departures from a continuously adjustable state of quasi-equilibrium (within specific extreme limits) in the internal-external water balance which occur as a consequence of changes in weather and the microclimate. The resting state for various behavioural patterns is induced by changes in meteorological conditions that produce hydration in relation to a previously adapted state of hygrothermal equilibrium for activity. Conversely, various states of activity involving locomotion and flight are susceptible to environmental stimulation when meteorological conditions exert a desiccating influence on a previously adapted hygrothermal equilibrium for the resting state (Fig. 1).

Although the quasi-equilibrium state for internal-external water balance may vary with physiological adaptation to new environments given sufficient time for adjustment, the reciprocating and alternating states of activity and rest are limited to a certain range of hygrothermal conditions for each species (Haufe, 1963c). Consequently, extremes in desiccation and hydration determine the duration of favourable activities that are essential in dispersal of populations, searching for hosts among animals and man, and successful completion of the life cycle. Hygrothermal relations in the environments selected by man and animals are of primary importance therefore in regulating or limiting the transmission and incubation of pathogens and the incidence of diseases associated with arthropod vectors.

Hygrothermal regulation of certain activities such as flight and locomotion may involve several meteorological factors interacting within a natural habitat. Temperature, solar radiation and water-vapor pressure are the primary elements that determine the exchange of water between the insect and its immediate environ-

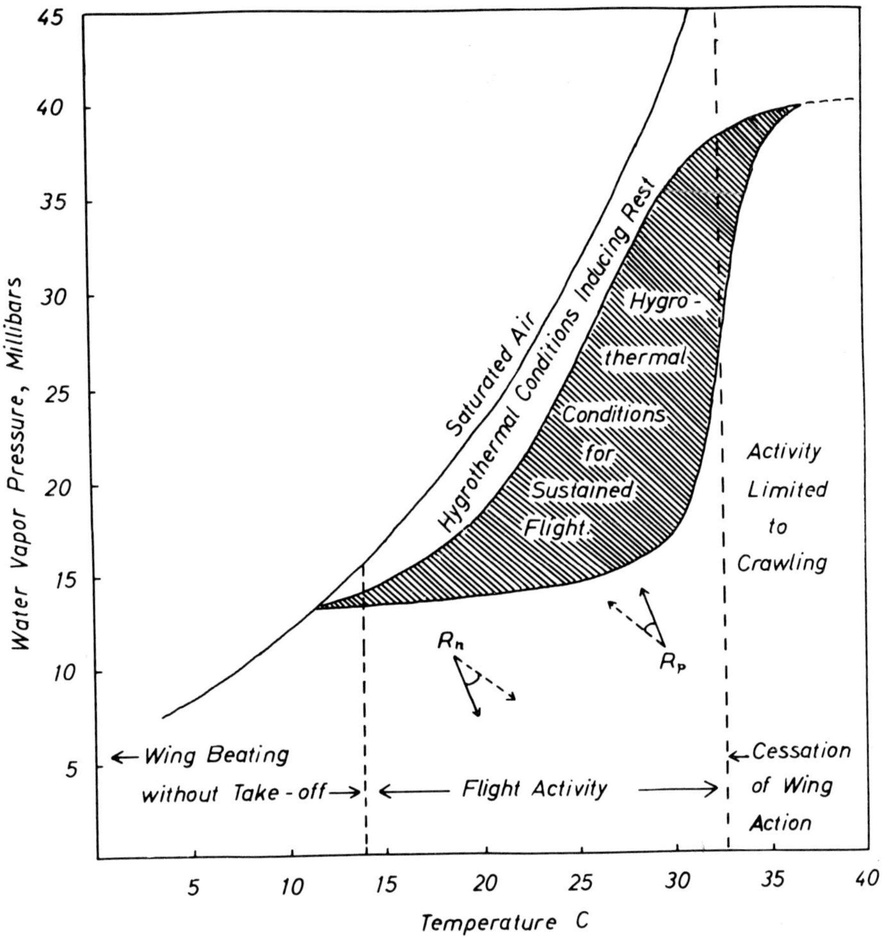

Fig. 1. Relation of rest and stimulated activities in the mosquito *Aedes aegypti* (L.) to hygrothermal conditions in a psychrometric diagram. Moments for change in environment producing desiccation (negative water balance) and hydration (positive water balance) designated by R_n and R_p, respectively.

ment. Water balance under a given set of these primary conditions is also influenced by wind in flying insects and by convective currents in resting insects (Haufe, 1957, 1963a).Synoptic correlations in subarctic environments between weather systems and flight activity (Haufe, 1957) indicate that the influence of atmospheric pressure may be due at least in part to changes in vapor pressure concomitant with compression and decompression in air masses.

It is especially significant that the dominant role of hygrothermal regulation in maintaining water balance within certain limits is also consistent with what is known of energy metabolism and its relation to alternating states of activity and rest in mosquitoes (Haufe, 1966). Water is the essential life-preserving medium supporting vital metabolic pathways for glycolysis in most insects (Fig. 2). Glycogen reserves are depleted during periods of sustained or intense activity and are restored during periods of reduced activity and rest (Harwood, 1964; Haufe, 1966; Rowley et al., 1968; Rowley, 1970). Exchange of water between the internal and external environments appears to be continuous in both directions. Net water balance, influenced by the hygrothermal environment, has a vital role in the direction of chemical processes in anaerobiosis. Furthermore, the direction of chemical processes associated with negative and positive changes in water balance between the internal and external environments corresponds directly with reactions of mosquitoes to desiccating and hydrating conditions associated with meteorological changes in the atmosphere. The correspondence between these behavioural and chemical systems in arthropods is in accord not only with anaerobiosis involving glycolysis but also with the metabolic flow of water in the aerobic degradation of carbohydrates (Fig. 3) as manifest in the tricarboxylic acid cycle (Haufe, 1966). According to recently integrated biometeorological studies, therefore, hygrothermal elements are the primary meteorological variables regulating the physiological interdependence of energy metabolism, internal-external water balance and the development and activities of mosquitoes.

The influence of humidity on development and behaviour has been reported for various arthropods of veterinary-medical importance. Activity of *Culicoides* spp. has a high positive correlation with humidity inland (Reuben, 1963) but not, apparently, in a marine climate (Kettle, 1969) with hygrothermal stability. Embryonic longevity is restricted within narrow ranges of humidity and temperature in mosquitoes (Meola, 1964) and in mites (Palimpsestov et al., 1960), and hatching percentage is directly related to humidity (Mulla et al., 1968). Flight activity of mosquitoes corresponds with hours of the day when changes in temperature and humidity are most pronounced (Bast, 1961). Flight in search of warm-blooded hosts is directed mainly by warm moist convection currents (Wright et al., 1962; Cumming et al., 1967) and differences in attraction of various surfaces may be related to the rate at which water evaporates from them. Wharton (1963) described the importance of equilibrium humidities for ticks and other organisms when water is not available. Hygrothermal conditions have an important influence on oögenesis and oviposition in ticks and account for the seasonal changes that occur in blood-engorged females as a result of the effect of water balance on the maturation of the gonads (Belozerov et al., 1966). Permeability of the integument of ticks was estimated to decrease by 4-5 times of the initial value during the course of engorgement of females on mammalian blood (Belozerov, 1967). Acitivity of ticks (Dyk et al., 1968) and aeronautic behaviour of spiders (Richter, 1970) occur during desiccating phases of a changing hygrothermal environment similar to those described for mosquitoes (Haufe, 1964). Guidicelli (1966) found that humidity was the

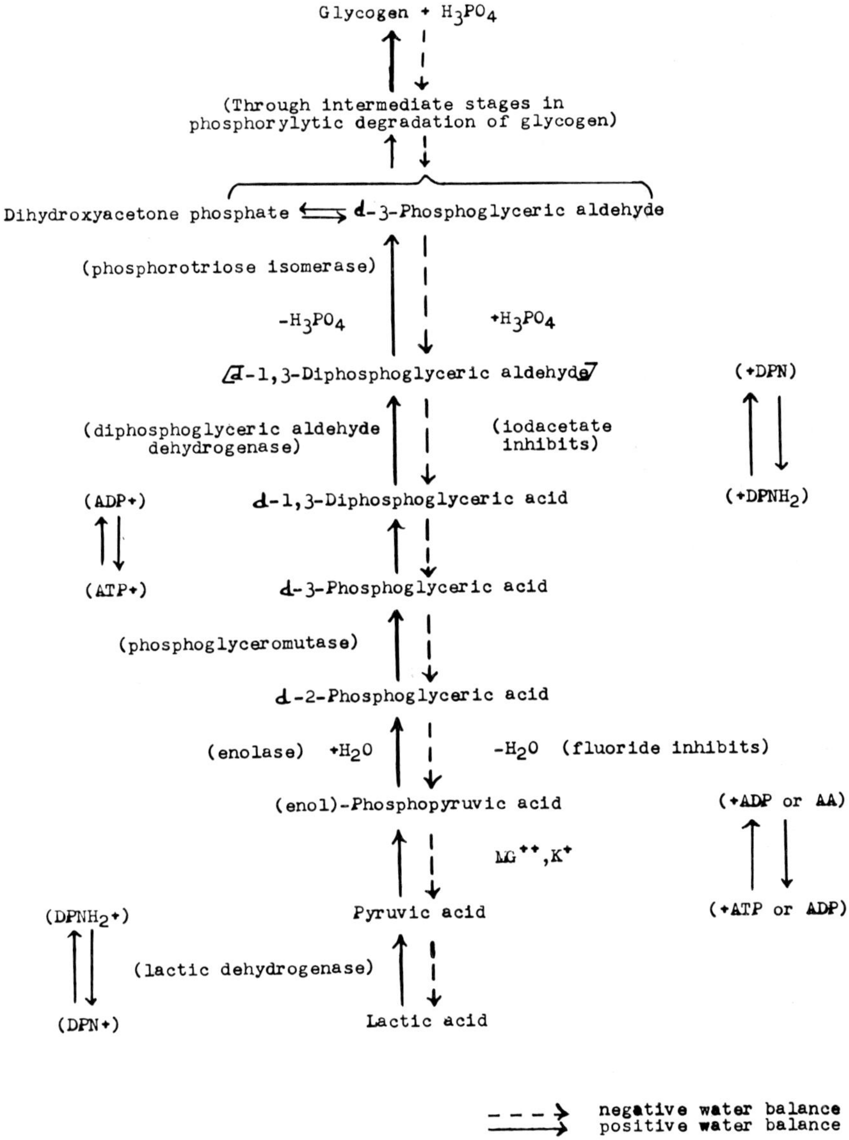

Fig. 2. Metabolic flow of water in processes of glycolysis in insects.

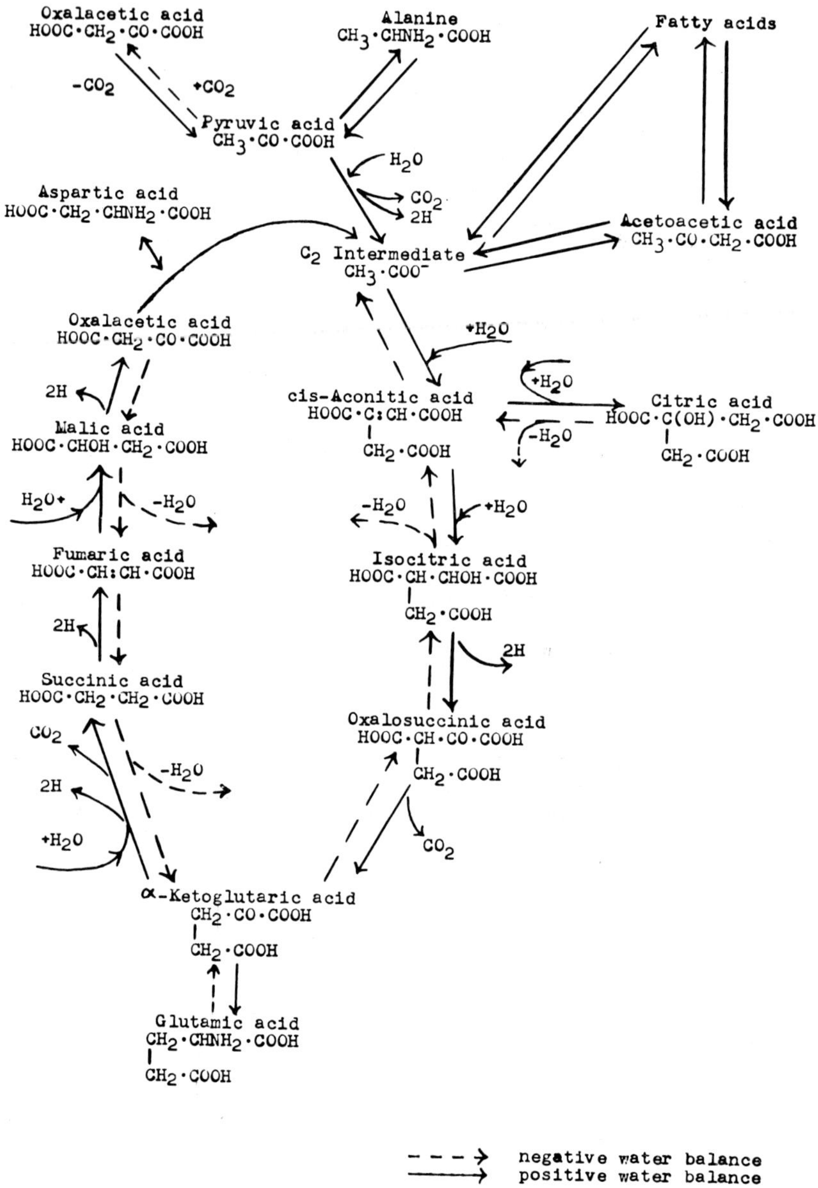

Fig. 3. Metabolic flow of water in processes of aerobic degradation of carbohydrates (tricarboxylic acid cycle) in insects.

primary factor in influencing biting activity of simuliids in tropical forest and savanna.

Some arthropods can take up water vapor from the air and retain a water balance well above a certain external relative humidity known as the equilibrium humidity. Within the first 24 hours after dehydration, tick larvae absorb water at lower humidities than those at which the equilibrium weight is maintained (Knülle, 1966). Adult fleas gain water from air at 82% R.H. in the first 24 hours after emergence (Humphries, 1967). Since there is no known physical process to explain these phenomena, which also occur in cockroaches and locusts, there is speculation that an active pumping mechanism exists in the integument of some arthropods (Winston, 1967). Significant differences in dimensions of the spiracles in members of the same sex and species of sandflies (Phlebotominae) from different districts were attributed to adaptation to varying conditions of humidity within colonies of the large gerbil serving as host (Dergacheva, 1967).

3. CIRCADIAN RHYTHMS AND STIMULUS-RESPONSE SYSTEMS

Circadian rhythms are apparent in the activity of most insects, especially in the flight and blood-feeding behaviour of most airborne vectors of disease. It is in these phenomena that several meteorological factors such as light, solar radiation, atmospheric pressure and electrical state of the atmosphere exert a triggering influence on patterns of activity within the limits of hygrothermal regulation (Haufe, 1962, 1963b, c, 1964b). Light, especially when expressed as the "crep" unit for duration of twilight (Nielson, 1961), appears to account for the swarming and biting activity of many crepuscular species of mosquitoes (Haddow, 1964; Haddow et al., 1968). Some studies (Dow et al., 1970) have shown that the day-to-day differences in catches of mosquitoes during diurnal cycles of activity are closely related to the relative humidity one hour after sunset. Stimulation of activity by changes in light and other "triggering" physical factors is related to rate of adaptation to hygrothermal changes (Haufe, 1961, 1962).

There is evidence that sensory-stimulating environmental factors interact with hygrothermal regulatory systems to govern the periodicity and rhythm of activity in arthropods other than mosquitoes. The number of active tsetse flies, *Glossina longipennis,* increases as saturation deficit increases until a low threshold of light is reached (Power, 1964). Norris (1966) found that activity tends to be bimodal on hot days and unimodal on cooler days for several species of blowflies, and temperature is always higher when activity ceases than when it began on the same day.

4. DISPERSAL, DISTRIBUTION AND VECTOR RELATIONS

Dispersal and distribution of arthropods are of major economic concern in pestilence and vector transmission of disease. Although migratory aspects of dispersal of insects by flight was recently reviewed comprehensively for some of the more frequently studied species (Johnson, 1969), there is still little significant progress in accounting for displacement and distribution of vectors in terms of predictable or manageable systems. Vectorial capacity in disease transmission has been assessed in biotic terms (Garrett-Jones, 1964) but these parameters have limited application in predicting disease outbreaks from biometeorological and other environmental information. Displacement activities and resulting epidemic periods may be associated with different critical physiological functions in different vectors, for example, oviposition in autogenous female simuliids (Fredeen, 1963), blood-feeding in certain vectors of encephalitis (Eldridge, 1963), or oögenesis in ticks (Belozerov et al., 1966). Distribution and numbers of vectors usually depend on a complex association of vegetation, climate, edaphic conditions and human or animal hosts in the environment (Morris 1962; Varma, 1965), and they are not affected appreciably by the occasional movements of certain host foci such as normal game species or humans (Foster, 1964).

Biometeorological aspects of dispersal and distribution of vectors in relation to disease are complex and require a much greater emphasis on ethological systems related to interacting populations of arthropods and host animals.

5. SYNOPTIC RELATION OF WEATHER TO VECTOR ABUNDANCE AND EPIDEMIOLOGY

Developments in synoptic approaches to biometeorology provide some measure of capability in understanding and exploiting fundamental biological reaction to environment as well as in studying the relations between weather and organisms. Little progress has been made on the earlier studies since recent reviews (Haufe, 1963a, b).

A synoptic model for predicting periodic variations in activity accords with the concept of hygrothermal regulation as a limiting system for mosquitoes in subarctic and temperate environments (Haufe, 1966a). Investigations on *Anopheles* spp. (Kühlhorn, 1963) have also confirmed the view that unfavourable weather is more limiting than biotic factors, especially over short periods. Epidemics of disease may be held in check by climate when optimal conditions for pathogenic parasites are also density-suppressing at the same time for the vectors (Nanda et. al., 1962). On the other hand, weather systems have been responsible for epidemics of vector-borne diseases (Aragão, 1964; Hayes et al., 1964) and for invasions of disease vectors and other arthropods far from foci of infection or breeding sources of populations (Garrett-Jones, 1962; Hurst, 1969).

There is increasing evidence recently to show that the electrical state of the atmosphere, particularly air ions, affects biological processes and insect activities (Haine, 1961, 1967; Maw, 1965). Negative ions promote moulting and adult emergence in some species. Robinson and Dirnfeld (1963) have related changes in the ionization state of the atmosphere to biologically important meteorological elements such as wind, solar radiation, temperature, and humidity. Variations in small-ion ratios associated with cold weather fronts may stimulate the emergence of moth (Helson et al., 1970). More elaborate synoptic studies are necessary in which meteorological factors serving as "triggering" mechanisms are integrated with hygrothermal models of weather systems to predict activities and geographic displacement of airborne arthropods.

Diapause and related forms of arrested development in insects is another phenomenon with scientific and practical significance in synoptic biometeorological studies. It ensures survival of the species during unfavourable weather conditions and is a critical link in the life cycle of insect vectors, especially in temperate and subarctic regions (Haufe, 1963b). An arrested ovarian development induced by a combination of low temperature and low humidity is an essential mechanism, even for tropical mosquitoes to continue the life cycle through periods of drought (Omer et al., 1968). Diapause in animal parasites and vectors may also be linked more critically with the host (Haufe, 1963b). For example, ultraviolet radiation combines with photoperiod during seasons of the year to induce changes in the constitution of bovine blood. Female horn flies, which are host-specific obligate parasites of cattle, are nutritionally sensitive to the blood changes and produce progeny with increasing predisposition to diapause in the pupal stage when fed on cattle exposed to decreased ultraviolet radiation (Depner, 1962). Phenomena of various forms of arrested development have special significance in host-parasite and vector relations when linked with meteorological factors since they represent critical functions synchronizing the life cycle with weather and climate. Consequently, in addition to their value in schemes to predict infestations, they also have potential use in environmental control measures.

REFERENCES: see pp. 148-151.

Laboratory Animals

by

W.H. Weihe

1. INTRODUCTION

The living conditons and the requirements of laboratory animals are dealt with in the field of laboratory animal science. The emergence of this field was instigated by the need for a healthy, standardized animal for experimentation. The laboratory animal is a sensitive biological system that should be in a constant state of high susceptibility and reproducibility of response. This can only be achieved by the standardization of all the factors of its ecosystem. Among these factors those of the physical environment have been ignored or could be only incompletely controlled for a long time. However, after the striking effects of extreme meteorological conditions on caged animals became known through intensified research in environmental physiology, great efforts for the elaboration of standardization measures were made.

Since 1960 systematic studies have been carried out on the most suitable conditions of the physical environment for the large scale breeding and maintaining of animals for research. In a survey under the title "The environment of laboratory animals" (Weihe, 1964) existing knowledge was compiled to demonstrate how the laboratory animal has to adapt to environmental conditions and their changes. Three types of climate were differentiated: (1) the macroclimate outside, (2) the room climate, and (3) the cage climate or cryptoclimate. The effective meteorological factors decrease in number, intensity and variation from outdoors to inside the cage. While in conventional animal facilities the variation of temperature, humidity, radiation, air turbulence and pollution was rather wide and more or less strongly affected the various animal species inside the cage, in modern fully air conditioned animal houses the situation has much changed. During the last 10 years an impressive number of high quality buildings has been erected with partial or full air conditioning systems which guarantee constant conditions throughout the year. With such buildings the standardization of the physical environment has now become possible. However, the actual conditions that the breeder and the researcher should provide for the animal in general or as a particular model are still under discussion.

The principal biometeorological problem in laboratory animal science is what physical conditions should be provided for given animal species, age groups, caging conditions, nutrition and other ecological factors for the particular purpose of the experiment. In order to find the proper answer for a given experiment comparative studies are carried out at various temperatures, and on the basis of the results the optimal conditions are chosen. The term "optimal" does not necessarily refer to the most favorable physiological conditions or the comfort for the animal but applies solely to the purpose for which the animal is being used. As the equipment for the performance of such comparative studies is still rare not many experiments of this kind have been carried out. A simpler approach has also been tried in providing as constant a physical environment as possible to which all species and experimental models are forced to conform. This simplification stems from the fact that for most researchers the effects of the factors of the physical environment are still considered to be insignificant.

At the present state of development of biometeorology in the art of animal experimentation 3 principal aspects can be selected for consideration:

1. The conditions of the physical environment of laboratory animals for breeding and basic experiments,
2. the permissible variation in the physical environment,
3. the provision of extreme conditions for the stimulation of selected functions.

2.THE CONDITIONS OF THE PHYSICAL ENVIRONMENT OF LABORATORY ANIMALS FOR BREEDING AND BASIC EXPERIMENTS

The vast majority of laboratory animals (mice, rats, hamsters, guinea pigs and rabbits) should live under physical conditions at which the reproduction rate is high, they experience the least strain and no particular acclimatization to daily or seasonal changes is required. This applies also to the less frequently used animal species for which no particular experimental conditions are required such as cats, dogs, sheep, and monkeys. The least strain is experienced if the thermal conditions are such that the active and feeding animal lives at thermoneutrality without metabolic or insulative adaptation.

3. SIGNIFICANCE OF TEMPERATURE AND HUMIDITY FOR LABORATORY ANIMALS

Thermoneutrality is synonymous with comfort temperature. Environmental temperature (Ta) is still the most important factor of the physical environment. Humidity will be considered as a factor insofar as its control becomes possible in modern animal houses. Early recommendations for room temperature and relative humidity were derived from empirical observations and data for human comfort and were not based on specific investigations. In these cases the different relative heat production and insulation of the animal species were not taken into account. Recommended environmental temperatures for mice and rats were between 20 and

22°C (Lane-Petter, 1963). For a long time this was considered to be too low for these two species. Therefore, Ta's were higher in the USA (24°C) than in England (18-20°C) (Weihe, 1965, a,b).

Recently Weihe (1971a) used the model of the increasing heat production of animals at temperatures below and above the thermoneutral zone of Scholander et al. (1950) to explain the different recommendations. For this the lower critical temperature of the thermoneutral zone, the minimum oxygen consumption (MOC) of the resting animal, and the steapness of the gradient of oxygen consumption versus environmental temperature were used to elaborate comfort conditions (Fig.1.). The better the insulation of the animal species the lower will be the lower critical temperature and the flatter the oxygen consumption gradient. Feeding and activity, even lactation increases the metabolic rate in average to about 30 to 50% above MOC values (Pennycuik, 1967a, b; Weihe, 1971a). Consequently with increased oxygen consumption from the physiological activities of the animal under the limited conditions of the cage, animals will maintain thermoneutrality at environmental temperatures below the lower critical temperature. The range between the lower critical temperature for the fasting and resting animal (MOC) and for the feeding and active animal depends on the steapness of the oxygen consumption or heat production gradient, being narrow for animals with poor body insulation such

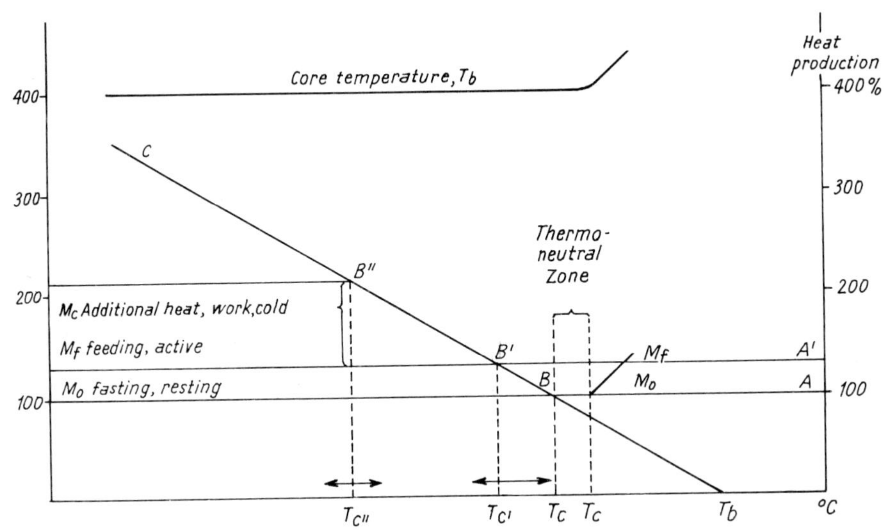

Fig. 1. Diagram of the heat production (oxygen consumption)vs. environmental temperature gradient. Minimum oxygen consumption is indicated as 100% heat production of M_o-line. The lower critical temperature (T_c) is where the gradient intersects the M_o-line. The feeding and active animal shows a higher heat production at M_f (+30%=130%) causing a shift of T_c to a lower T_a at T_c. Data for the common animal species based on the inclination of the gradient are given in Table 1 (from Weihe, 1971a).

as the hairless mouse and wide for animals with good insulation such as the rabbit (Mount, 1971; Pennycuik, 1967a, b; Weihe, 1971a). Data are given in Table 1. For some species such as the guinea pig these values appear rather high if compared with those previously given (Lane-Petter, 1963). This is due to the fact that they refer to single animals in cages without insulative protection. The values will move down to lower environment temperatures if the animals are placed in groups and provided with bedding material in cages with solid walls to allow for behavioral temperature regulation as in nature. The animal will make every effort to avoid metabolic adaptation (Cabanac, 1972; Ogilvie, 1970; Prychodko, 1958, Weihe, 1971b).

A good example of this is the mouse, the smallest and most frequently used animal species for research. Single mice in solid wall cages with bedding will not show a signigicant increase of their energy requirement until below 22°C and 2 or 6 mice until below 20°C (Weihe, 1971b). Therefore, as long as even such small animals as mice are kept in groups in cages which allow for insulation against the ambient conditions, environmental temperatures down to 20°C are justified for maintenance as there will be no significant degree of metabolic adaptation. Indeed, this is in agreement with the early empirical recommendations and the comfort temperatures for man. However, only since the great importance of behavioral temperature regulation has been recognized the recommendations became meaningful. If the animals are kept in wire mesh cages, where they have little chance to improve their insulation beyond autonomic physiological responses, higher temperatures (Table 1) should be provided.

Table 1. Recommended comfort temperatures for minimum metabolic adaptation of the single animal in cages with free air exchange (after Weihe, 1971a).

Species	Comfort Range (°C)	Cooling Power Dry Kata[a] (mcal cm $^{-2}$ sec $^{-1}$)
Hairless mouse	31-34	1.5 - 0.9
White mouse	26-31	2.7 - 1.5
Guinea pig	24-29	3.3 - 2.0
Rat	24-29	3.3 - 2.0
Hamster	21-26	4.1 - 2.7
Rabbit	15-21	5.7 - 4.1

All these data refer to conventional or SPF – specified pathogen free – laboratory animals. Increasing numbers of the same species are being used as germfree animals which are a new biological entity because of the lack of viruses, bacteria etc. Recently the MOC of germfree rats was found to be about 30% lower than that of conventional or SPF animals (Levenson et al, 1969). Germfree animals then, require higher ambient temperatures than their counterparts harboring bacterial flora.

4. SIGNIFICANCE OF AIR EXCHANGE

Under conditions of poor ventilation the insulative value of a cage can be higher than is desirable so that the thermal conditions inside and outside the cage are different and a definite cage climate is building up. Such differences were reported by Weihe (1965a) and Murakami (1971). They are always the result of a poor air exchange between the room and cage air in cages with either too high walls or lids with inadequate perforations or in cages placed in remote parts of the room away from the air stream for ventilation. Extreme differences were found when cages were covered with filter caps to prevent contamination by airborne bacteria, as is common practice. Within a few hours the cage temperature rose to 22.2°C compared to the room temperature of 19.2°C and the relative humidity increased from 40 to 50%. At a higher room temperature of 23.3°C the cage temperature rose less to 25.5°C and the relative humidity increased from 70 to 75% (Simmons et al., 1968). In these experiments fresh shavings were used. If shavings are left in the cage for days so that they are saturated with urin the humidity in badly ventilated cages will increase and can rise to almost saturation point in the air layers at the level of the animals head, 2-5cm above the ground. Modern cages are built and room ventilation is designed in such a way that there is little danger of the build up of an unwanted cage climate.

It has been a problem for some time to avoid overheating of rooms when occupied with large numbers of different species of animals. In particular the ventilation engineer must know the heat production of animals under various conditions of caging to calculate the required number of air exchanges per hour for a room. Much confusion has existed about the average 24-hr oxygen consumption of laboratory animals from which the heat production is calculated. Brewer (1964) gave values as high as 2-3 fold MOC values. Such increases are observed only over short intervals of minutes with special exercising equipment provided or at room temperatures near zero°C in animals with high metabolic adaptation. Weihe (1971a) recommended a 30 to 50% increase above MOC values as a mean for one or 24 hours, based on metabolic body size ($kg^{0.75}$) as described by Kleiber (1961). This has proved to be a good basis from which the required hourly ventilation rate for an animal room can be calculated to avoid heat accumulation. Based on Kleiber's factor a diagram has been devised by Weihe (Gesellschaft für Versuchstierkunde, 1971) from which the heat production of the physiologically active animal under thermoneutral conditions based on body weight can be found for any animal species and weight (Fig. 2).

5. SIGNIFICANCE OF AIR MOVEMENT

Air flow has always been ignored in the animal house. Drafts on the one hand and pockets of still air on the other hand were a common phenomenon not only in conventional but also in air conditioned rooms. This gave rise to much variance in the response of animals in an experiment and facilitated the spreading of infections.

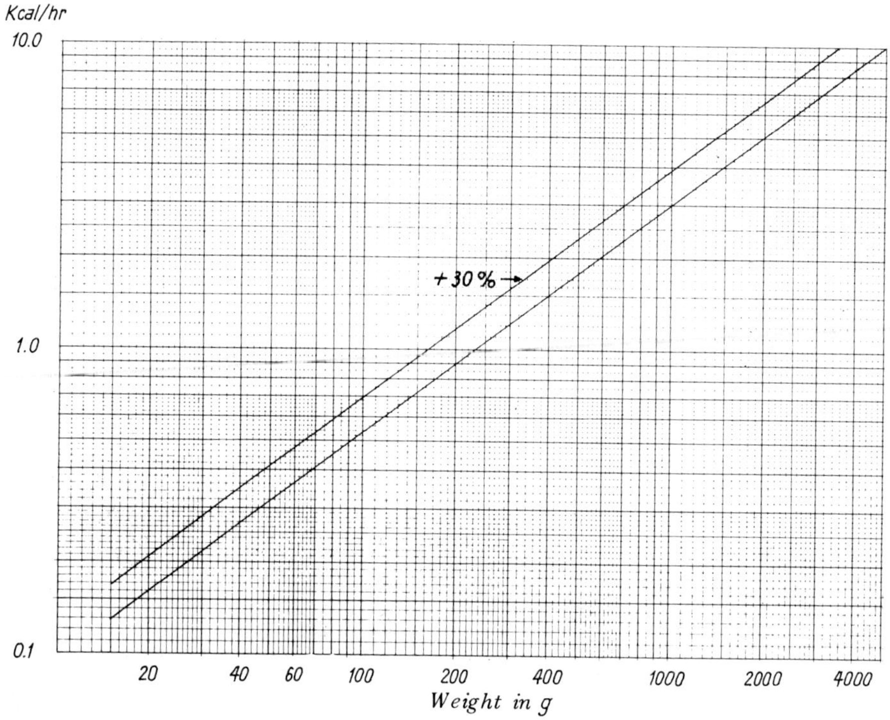

Fig. 2. Relationship between body weight and heat production based on the metabolic body size ($W^{0.75}$) acoording to Kleiber (1961) for the estimation of the mean heat production of laboratory animals under thermal conditions of minimum metabolic adaptation. Lower line Kleiber's values for the inactive, fasting animal, upper line (+30%) for the active and feeding animal (from Weihe, 1971b).

The systematic investigation of the effects of air flow as a biotropic factor began with the introduction of a new technique for housing animals such as the placing of cages in a sterile horizontal laminar air flow with 45 cm/sec and higher air velocity (De Koning et al., 1970). Such high air velocities are necessary to protect animals from contact with contaminated air outside the sterile air stream. The question arose as to wether the exposure in the air stream would be harmful to the animal causing irritation or strong cooling. The convective cooling effect can be well demonstrated in fully exposed mice in wire mesh cages and increases at constant Ta linearly with the square root of the air velocity (cm/sec) (Weihe, 1970). This cooling effect from increased convective heat loss can be eliminated by increasing Ta of the air stream. When animals, however, are placed in cages with solid walls this will not even be necessary because they will use behavioral temperature regulation and take shelter in the still air areas of the cage (Weihe, unpublished data). This is a further example of the strong compensatory capacity of behavioral temperature regulation of the animal if means are provided for it.

It has now become routine to take three factors of the physical environment of the animal room into consideration: dry bulb temperature (Ta), relative humidity (rh), and air velocity (v). Modern animal facilities for all small animal species should provide a constant room temperature at 22°C with a variation of less than 2°C, a humidity of 50% rh with a variation of less than 10% and an equal air flow (v ~ 5 cm/sec) through all parts of the room. The number of air exchanges should be no less than 10 exchanges/hr to avoid heat accuumulation at any time (Gesellschaft für Versuchstierkunde, 1971). The avoidance of cold and heat radiation has always been a technical problem for the architect but has not been given much attention as yet in animal houses.

6. THE PERMISSIBLE VARIATION IN THE PHYSICAL ENVIRONMENT

The permissible variation of environmental temperature at the limits of the thermoneutral zone, for the active and feeding animal, or a zone decided on for the particular experimental purpose, without stimulation of major adaptive response or stress of the animal, is still a matter of debate. For breeding colonies and basic experiments Weihe (1970) recommended an environmental temperature range of 15 to 32°C within which the animals (mouse, rat, hamster, guinea pig) in solid wall cages with bedding will respond with only slight metabolic adaptation without major strain (Fig. 3). This statement was based on a number of investigations in the literature. Also humidity can vary within the range of 30-80% rh for the given environmental temperature (Weihe, 1965a, b).

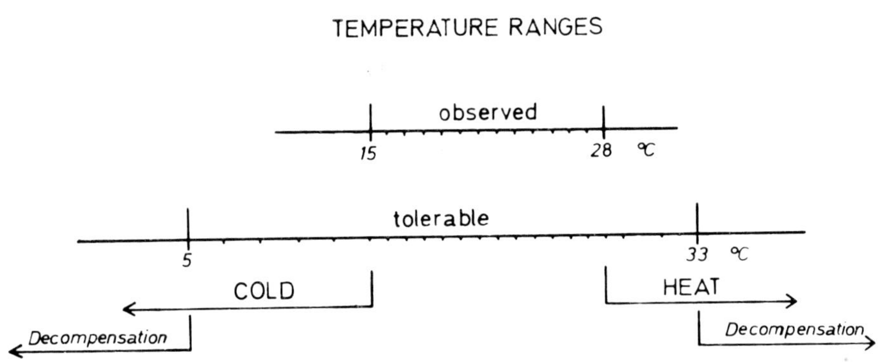

Fig. 3. Temperature range to demonstrate the zone of minimum metabolic adaptation observed for mice, rats, hamsters and guinea pigs in cages with shavings, the zones of moderate cold and heat adaptation and the zones of extreme cold and heat adaptation with increasing signs of decompensation of homeostasis (from Weihe, 1970).

Variations within these ranges are usually noticed in conventional animal rooms through outside weather changes and are reflected in the daily food and water intake and body weight of the animals. The correlation of the daily body weight of rats with temperature and humidity changes outside or inside a conventional animal room revealed a highly significant correlation with increase of body weight after cool nights and decrease of morning body weight after warm nights (Weihe et al., 1961). Also the acitivity of mice (Stille, Brezowsky and Weihe, 1968) and the behaviour of cats (Tolcsvai-Nagy et al,. 1968) in a conventional laboratory were weather and season dependent.

Several investigations were carried out in which groups of mice and rats were raised at 2 or 3 different constant temperatures and compared with groups which were changed at 12 to 24-hr intervals between such temperatures (Anderson et al., 1968; Harrison et al., 1964; Knudson, 1962; Moore, 1944; Roubicek et al., 1968). The temperatures were as high as 35°C, which is near the limits of tolerance. Taking the fertility, reproduction and growth of litters of mice and rats as parameters, it was found that these changes had no inhibitory effect if they were arranged in a physiological pattern, such as a high temperature during the day and a lower temperature during the night. However, all experiments have shown that major variations of temperature caused a moderate or significant reduction of growth and development (Harrison et al., 1969).

Such experiments on the reproduction and growth of rats in breeding colonies and maintaining of animals of different ages were used by Weihe (1965a, b) to work out temperature-humidity climatograms. From such climatograms the permissible variance of T_a and rh for breeding and maintaining animals can be read.

From the available knowledge it is concluded that the weaker and more vulnerable the animal the smaller the variation of a given environmental temperature and relative humidity should be. This applies to young and old animals and those under nutritional or experimental stress. The stronger the stress during the experiment and the more constant the responses of an animal in the experiment are required to be, the more constant the thermal conditions in the environment should be. In any case the animal will have to adapt physiologically and behaviorally to regular and irregular changes of the ambient thermal conditions. Such adaptation responses which may be a stimulans for the preexperimental healthy strong animal can become a severe strain for the weakened animal during the experiment and can terminate the experiment prematurely. Less strict thermal constancy can be permitted in breeding quarters of laboratory animals where a moderate variation of thermal conditions seems to activate development (Weihe, 1965a).

7.THE PROVISION OF EXTREME CONDITIONS FOR SPECIAL PURPOSES

For many experimental purposes the stimulation of a particular physiological function of the whole animal is indicated. In such cases the animals are exposed for short or long intervals continuously or intermittently to extreme conditions of heat, cold or altitude. For example a high metabolic rate can easily be achieved by cold exposure and removal of all means for insulation in wire mesh cages.

Extreme conditions are usually near the limits of tolerance, where, because of stress homeostasis is near decompensation and animals become highly sensitive (see Fig.3). Therefore knowledge of the limits of tolerance is mandatory for the planning of experiments with stimulation of selected functions. A vast number of studies have been carried out on the determination of the limits of tolerance and the failing of the various functions for which the limits are defined. Such studies belong to the field of environmental physiology and will be reported elsewhere in this volume. Mice can be bred at as low a temperature as -3°C if enough material for insulation is provided (Barnett, 1965) and at as high a temperature as 36°C (Pennycuik, 1964; 1967a, b; 1969) and reproduction of mice and rats seems to be possible up to 4000 m altitude (Weihe, 1965b). It has become clear that at these limits of temperature and altitude and beyond them ovulation, size of litters, and lactation are seriously hampered and length of gestation altered. Heat and altitude exposure of animals have proved to be especially valuable for research in teratology (Pennycuik, 1969). Strain differences also become apparant only near the limits of tolerance and are important for sorting out and selective breeding of heat or cold resistant mice and rats (Barnett and Manly, 1959; Barnett et al., 1959; Ogilvie, 1970). The ontogenesis of animals is different in the cold than in the heat. Tails of mice and rats growing in the cold are short and thick and long and slim in the heat (Harrison et al, 1960; Moore, 1944). Such traits may partly disappear later in life but can be of practical importance simply because the intravenous injection into the tail vein is easier in heat exposed animals. The thickness of the fur changes with long heat or cold exposure (Barnett, 1959). This is of importance in the study of cold-exposed animals when the increased metabolic activity decreases because of improved insulation by the fur or for particular studies of the skin and hair growth. The body fat reserves and the subcutaneous fat layer of cold exposed animals is reduced compared with thermoneutral animals and the composition of reserve fat is altered (Harland et al., 1969). This may prove to be important for specific metabolic studies. The shortage of the oxygen at high altitudes leads to a depression of metabolic rate with a reduction of energy intake and reduced growth during the adaptation period, which is short in young and long in old animals (Weihe, 1966). This applies to altitudes of about 3.500 m. Exposure of mice and rats at extreme altitudes such as 6000 m is common practice to stimulate erythropoesis.

With the increased metabolic rate in cold-exposed animals the toxicity of drugs can be changed. Increases in toxicity of drugs by a factor of 1000 have been observed in animals acutely exposed to heat and cold. To investigate the effect of a drug on the metabolism or on functions controlling heat loss or on temperature regulation, animals have to be exposed to temperatures where an increase of metabolic rate can be ensured. Cold-exposed animals are the only tool for such studies. However, it is not justified to infer from such studies directly to man as metabolic adaptation, which is easily induced in laboratory animals, is avoided by man at all events (Weihe, 1972). For studies on learning in experimental psychology or testing of drugs affecting learning rats raised in the cold (12°C) proved better in learning than rats raised in the heat (32°C) (Moore, 1944).

8. THE SIGNIFICANCE OF NON-THERMAL FACTORS IN THE PHYSICAL ENVIRONMENT

A number of studies have been carried out on the effects on animals of weak environmental factors such as electromagnetic fields and air ions. Some experiments are reported in this volume. Up to now such experiments have shown evidence that under fully controlled conditons of the thermal environment such factors do affect the animals. However, it is too early to consider the control of these factors to be important in breeding and housing experimental animals. More work has to be done and recording of these factors must be completed particularly in fully climatized modern animal houses before practical advice and directions can be outlined.

REFERENCES: see pp. 151-152.

REFERENCES*

CHAPTER 9

DIRECT EFFECTS OF THE PHYSICO - CHEMICAL ENVIRONMENT OF PLANTS AND TREES ON HUMAN LIFE

(W. BAIER and A. CHAN)

ALLEN, JR., L.H. (1968): The energy budget at the earth's surface: A study of the carbon dioxide concentration monitored over an agricultural field near Ithaca, N.Y. N.Y. State Coll. Agric. Cornell Univ., Ithaca. Techn. Rep. Ecom., 2-681-3., 254 pp.

AMERICAN ASSOCIATION OF NURSERYMEN (1970): Green survival and the environmental crisis. Amer. Ass. Nurserymen, 9 pp.

BAUMGARTNER, A. (1970): Sauerstoffumsätze von Bäumen und Wäldern. Allg. Forstz., 22: 482-483.

BERRY, C.R. and RIPPERTON, L.A. (1963): Ozone, a possible cause of white pine emergence tipburn. Phytopathol., 53: 552-557.

BLANCHER, G. (1963): Les espaces verts urbains. Rev. Hyg. Méd. soc., 11: 219-237.

CHAN-TZU-CHEY. (1959): Hygienic background for determining the area of the sanitary protection zones around gas-filling stations. Gigiena i Sanitariya, 24: 17-21.

CHEPIL, W.S. (1965): Transport of soil and snow by wind. Meteorol. Monogr., 6: 123-132.

FOY, C.L. and BINGHAM, S.W. (1969): Some research approaches toward minimizing herbicidal residues in the environment. Res. Rev., 29: 105.

FOY, C.L. (1970): Plants and Pollution. In: Agronomy and Health. Am. Soc. Agron., Spec. Publ. 16: 37-59.

HANSBROUGH, J.R. (1967): Air quality and forestry. In: Agriculture and the Quality of our Environment. N.C. Brady (e.d.). Amer. Ass. adv. Sci., Publ.: 85.

HEPTING, G.H. (1964): Damage to forests from air pollution. J. Forest., 62: 630-634.

HESS, W. and KÜRSTEINER, E. (1962): Grundsätzliches zur Dämmung des Verkehrslärms mit Hilfe von Grünpflanzungen. In: Wohltätiger Wald. Neue Aufgaben unserer Wälder. Schweiz. Forstverein. Zürich, Bühler Buchdruck: 56-64.

HILL, A.C., PACK, M.R., TRESHOW, M., DOWNS, R.J. and TRANSTRUM, L.G. (1961): Plant injury induced by ozone. Phytopathol., 51: 356-363.

HORN, K. (1963): Einige Bemerkungen zur Frage der Festlegung sogenannter „Hygienischer Schutzzonen" für Zwecke der Stadt- und Gebietsplanung. Angew. Meteorol., 4: 260-262.

KÄMPFER, M. (1967): Lärmdämpfung durch raumplanerische Massnahmen, Grünanlagen und Wald. Bibl. Bundesanst. Vegetationskunde, Naturschutz, Landschaftspfl., Bad Godesberg, 6: 10 pp.

KELLER, TH. (1968): Der Einfluss der Luftverunreinigungen auf den Wald im Lichte der neuesten Literatur. Schweiz. Z. Forstwesen, 4-5: 353-363.

KING, E. (1970): Ökologisch-meteorologische Untersuchungen an Windschutzstreifen im nassen Sommer 1968. Agr. Meteorol., 7: 235-253.

LEE, D.H.K. (1970): Environmental quality and the ISB. In: Biometeorology 4, S.W. Tromp and W.H. Weihe (eds.): 23-25.

---

* References with more than one author are classified according to first author and year of publication.

MUKAMMAL, E.I. (1965): Ozone as a cause of tobacco injury. Agr. Meteorol., 2: 145-165.
MUKAMMAL, E.I. (CHAIRMAN) (1968): Air pollutants, meteorology and plant injury. WMO, Geneva, Techn. Note 96, 73 pp.
NÄGELI, W. (1954): Die Windbremsung durch einen grösseren Waldkomplex (The braking effect on the wind of a large forest). Mitt. Schweiz. Anst. forstl. Versuchswesen 29: 2.
NÄGELI, W. (1965): Über die Windverhältnisse im Bereich gestaffelter Windschutzstreifen (On the wind conditions in the range of staggered shelterbelts. Model tests with reed mats in the open field). Mitt. Schweiz. Anst. forstl. Versuchswesen, 41: 219-300.
NATIONAL AIR POLLUTION CONTROL ADMINISTRATION (1970): Recognition of air pollution injury to vegetation: a pictorial atlas. Inform. Rep. No. 1 TTR-7 agricult. Comm., Nat. Air Poll. Contr. Admin. Raleigh, North Carolina.
OTTO, H.W. and DAINES, R.H. (1969): Plant injury by air pollutants: Influence of humidity on stomatal apertures and plant response to ozone. Science, 163: 1209-1210.
PANEL ON THE PLANT SCIENCES (1966): The plant sciences now and in the coming decade. Nat. acad. Sci., Nat. Res. Council, Wash. D.C. Publ. 1405, 167 pp.
RAYMOND V. (1963): Role des espaces verts dans la prevention des maladies infectieuses des voies respiratoires. Rev. Hyg. Méd. soc., 11: 239-242.
RICH, SAUL (1968): Plants as air purifiers. Frontiers Plant Sci., 21: 6-7.
RICH, S., WAGGONER, P.E. and TOMLINSON, H. (1970): Ozone uptake by bean leaves. Science, 169: 79-80.
ROBINETTE, GARY (1968): Plants as air conditioners. Horticulture, 46: 26-29.
ROGALINA, T.S. (1960): Green plantations for the territory of factories and plants. Priroda, 10: 46-50.
SCURFIELD, G. (1960): Air pollution and tree growth. Forest Abstr., 21: 339, 517.
SHEAR, G.M. (1968): The development of the no-tillage concept in the United States. Outlook Agr., 5: 247-251.
SKIDMORE, E.L., FISHER, P.S. and WOODRUFF, N.P. (1970): Wind erosion equation: Computer solution and application. Soil Sci. Amer. Proc., 34: 931-935.
SPECTOR, W.S. (ed.) (1956): Handbook of Biological Data. Saunders Co., Philadelphia/ London, 584 pp.
STAPLE, W.J. and LEHANE, J.J. (1955): The influence of field shelter belts on wind velocity, evaporation, soil moisture and crop yield. Canad. J. agr. Sci., 35: 440-453.
TURNER, H. (1968): Der heutige Stand der Forschung über den Einfluss des Waldes auf das Klima. Schweiz. Z. Forstwesen, 4-5: 335-352.
VAN EIMERN, J. (ed.) (1964): Windbreaks and Shelterbelts. WMO Geneva, Tech. Note 59, 188 pp.
VAN EIMERN, J. (1968): Problems of shelter planning. In: Unesco Nat. Resources Res., Paris, Agroclimatol. Methods, Proc. Reading Symp., VII: 157-166.
WAGGONER, P.E., FURNIVAL, G.M. and REIFSNYDER, W.E. (1969): Simulation of the microclimate in a forest. Forest Sci., 15:37-45.
WANG, J.Y. (1970): Introduction to Environmental Studies. (Human Ecological Problems). Milieu Inform. Serv., San José, Calif., 320 pp.
WENT, F.W. (1960): Blue hazes in the atmosphere. Nature, 187: 641-643.
WENT, F.W. (1966): Dispersion and disposal of organic materials in the atmosphere. Desert Res. Inst., Prepr. Ser. 31, Univ. Nevada, Reno, 10 pp.
YANISHEVA, N. YA. (1969): Hygienic background for determining sanitary protection zones around electric power stations operating on peat. Gigiena i Sanitariya 24: 6-10.

CHAPTER 10

METEOROLOGICAL EFFECTS ON ANIMALS IMPORTANT FOR MAN

Chapter 10, Section 1a: Milkproduction of Cattle (H.D. JOHNSON)

BATEMAN, J.V. and ALBA, J. de (1961): Relative protein content of milk produced by four breed groups in a tropical environment. J. Dairy Sci., 44: 1190.

BIANCA, W. (1965): Reviews of the progress of dairy science, Section A: Physiology. Cattle in a hot environment. J. Dairy Res., 32: 291-345.

BOND, T.W. (1961): Recent advances in animal microclimatology. In: Proc ASAE Winter Meeting, Chicago, Ill., December.

BRODY, S., RAGSDALE, A.C., THOMPSON, H.J. and WORSTELL, D.M. (1954a): The thermal effects of radiation intensity (light) on milk production, feed and water consumption, and body weight in Holstein, Jersey, and Brahman cows at air temperatures 45, 70 and 80° F. Monogr. agr exp. Stat. Res. Bull., 556: 3-20.

BRODY, S., RAGSDALE, A.C., THOMPSON, H.J. and WORSTELL, D.M. (1954b): The effect of wind on milk production, feed and water consumption, and body weight in dairy cattle. Monogr. agr. exp. Stat. Res. Bull., 556:3-20.

BRODY, S., RAGSDALE, A.C., YECK, R.G. and WORSTELL, D.M. (1955): Milk production, feed and water consumption, and body weight of Jersey and Holstein cows in relation to several diurnal temperature rhythms. Monogr. agr exp. Stat. Res. Bull., 578.

CARGILL, B.F., STEWART, R.E. and JOHNSON, H.D. (1962): Effect of humidity on total room heat and vapor dissipation of Holstein cows at 65, 80 and 90° F. Monogr. agr. exp. Stat. Res. Bull., 794.

COBBLE, J.W. and HERMAN, H.A. (1951): The influence of environmental temperature on the composition of milk of the dairy cow. Monogr. agr. exp. Stat. Res. Bull., 485.

DAVIS, H.P. (1914): Effect of open-shed housing as compared with the closed stable for milk cows. Rep. Pa. agr. exp. stat. Bull., 183.

DICE, J.R. (1935): The nutrients required by dairy cows kept in an open shed versus cows kept in a dairy barn. J. Dairy Sci., 18: 447-448.

DICE, J.R. (1940): The influence of stable temperature on production and feed requirements of dairy cows. J. Dairy Sci., 23: 61-69.

DICE, J.R. (1942): The ability of yearling heifers to withstand cold temperatures. J. Dairy Sci., 25: 678-699.

GACULA, M.C., GAUNT, S.N. and DAMON, R.A. (1968): Genetics and environmental parameters of milk constituents for fine breeds. J. Dairy Sci., 51: 428-444.

HAHN, G.L., JOHNSON, H.D., SHANKLIN, M.D. and KIBLER, H.H. (1965): Inspired air cooling for lactating dairy cows in a hot environment. Trans. ASAE, 8, 3: 332-337.

HAHN, G.L. (1969): Predicted versus measured production differences using summer air conditioning for lactating dairy cows. J. Dairy Sci., 52: 800-802.

HAHN, G.L. and OSBURN, D.D. (1969): Feasibility of summer environmental control for dairy cattle based on expected production losses. Trans. ASAE., 12: 448-451.

HAHN, G.L. (1970): Probability of occurrence of cloud cover during summer daytime hours. In: Proc. ASAE Winter Meeting: 1-13.

HAHN, G.L. and OSBURN, D.D. (1970): Feasibility of evaporative cooling for dairy cattle based on expected production losses. Trans. ASAE., 13: 289-294.

HEINEMANN, B. (1947): The relation of percent total solids in separated milk and percent fat in whole milk at atmospheric temperature. J. Dairy Sci., 30: 757-762.

HOUSTON, J. and HALE, R.W. (1932): The yield and composition of milk with special reference to the influence of climatic conditions and other factors. J. Dairy Res., 3: 294-309.

JOHNSON, H.D., KIBLER, H.H., RAGSDALE, A.C. and SHANKLIN, M.D. (1960): Effects of various combinations of temperature and humidities on milk production. J. Dairy Sci., 43: 871.

JOHNSON, H.D., HAHN, G.L., KIBLER, H.H. and MERILAN, C.P. (1962a): Heat acclimation effects on lactation and related physiological responses of cattle. J. anim. Sci., 21: 1025.

JOHNSON, H.D., RAGSDALE, A.C., BERRY, I.L. and SHANKLIN, M.D. (1962b): Effect of various temperature-humidity combinations on milk production of Holstein cattle. Monogr. agr. exp. Stat. Res. Bull., 791.

JOHNSON, H.D., RAGSDALE, A.C., BERRY, I.L. and SHANKLIN, M.D. (1963): Temperature-humidity effects including influence of acclimation on feed and water consumption of Holstein cattle. Monogr. agr. exp. Stat. Res. Bull., 846.

JOHNSON, H.D. (1965a): Response of Animals to heat. Meteor. Monogr., 6: 109-122.

JOHNSON, H.D. (1965b): Environment temperature and lactation (with special reference to cattle). Int. J. Biometeor., 9: 103-116.

JOHNSON, H.D. (1967): Climatic effects on physiology and productivity of cattle. Ground Level Climatol.: 189-206.

JOHNSON, H.D., HAHN, G.L., KIBLER, H.H. SHANKLIN, M.D. and EDMONDSON, J.E. (1967): Heat and acclimation influences on lactation of Holstein cattle. Monogr. agr. exp. Stat. Res. Bull., 916.

KALI, J., MORAG, M. and AMIR, S. (1968): Seasonal changes in milk production and fertility in high yielding dairy cows in a desert climate. Int. J. Biometeor., 12: 271-275.

KELLY, M.A.R. and RUPEL, I.W. (1937): Relation of stable environment to milk production. U.S. Dept. Agr. techn. Bull., 591.

KELLY, C.F., BOND, T.E. and HEITMAN, H. (1954): The role of thermal radiation in animal ecology. Ecology, 35: 562-569.

KING, J.O.L. (1955): Variations in the quantity and composition of milk yielded by diseased cows while their body temperatures were elevated. Vet. Rec., 67: 432.

LECKY, T.P. (1950): Cattle breeding in Jamaica. Conference on the Improvement of Livestock under Tropical Conditions. Edinburgh, 5 pp (mimeograph).

LEE, D.H.K. (1951): Plant and animal nutrition in relation to soil and climate factors. Specialist Conf. Agr., Melbourne, Australia. (London, H.M.S.O.).

MAHADEVAN, P. and GALUKANDE, E.B. (1962): A genetic study of the Sahiwal grading-up scheme in Kenya. Anim. Prod., 4: 337.

MAHADEVAN, P. and HUTCHISON, H.G. (1964): Performances of crosses of *Bos taurus* and *Bos indicus* cattle for milk production in the coastal region of Tanganyika. Anim. Prod., 6: 331-336.

MCDOWELL, R.E., MOODY, G.E., VAN SOEST, P.J., LEHMAN, R.P. and FORD, G.L. (1969): Effects of heat stress on energy and water utilization of lactating cows. J. Dairy Sci., 52: 188-194.

MCINTYRE, K.N. (1967): A comparative study of the performance of dairy cows during summer and winter in a subtropical climate. Aust. J. exp. Agr. anim. Husb., 7: 400-407.

MERILAN, C.P. and BOWER, K.W. (1959): Influence of daily environmental temperature cycles on composition of cows' milk. Monogr. agr. exp. Stat. Bull., 687.

MONTERO, R.E. (1966): Problems in milk production in the tropics. J.Dairy Sci., 49: 1303-1305.

NELSON, P.E., SCHUH, J.D. and STOTT, G.H. (1967): Influence of season on leucocytes in milk. J. Dairy Sci., 50: 978.

PAYNE, W.J.A. (1957): The problem of milk production in tropical countries. Trop. Agr., 33: 34.

PAYNE, W.J.A. and HANCOCK, J. (1957): The direct effect of tropical climate on performance of European-type cattle. Empire. J. exp. Agr., 25: 321-338.

POPOFF, J.S. (1927): Der Einfluss der Dalte auf die Produktivität der Milchkühe. Z. Tierz. Zücht. Biol., 10: 285-291.

RAGSDALE, A.C., BRODY, S., THOMPSON, H.J. and WORSTELL, D.M. (1948): Influence of temperature, 50° to 105°F, on milk production and feed consumption in dairy cattle. Monogr. agr. exp. Stat. Res. Bull., 425.

RAGSDALE, A.C., WORSTELL, D.M., THOMPSON, H.J. and BRODY, S. (1949): Influence of temperature, 50° to 0°F and 50° to 95°F on milk production, feed and water consumption and body weight in Jersey and Holstein cows. Monogr. agr. exp. Stat. Res. Bull., 449.
RAGSDALE, A.C., THOMPSON, H.J., WORSTELL, D.M. and BRODY, S. (1950): Milk production and feed and water consumption responses of Brahman, Jersey and Holstein cows to changes in temperature 50° to 105°F and 50° to 8°F. Monogr. agr. exp. Stat. Res. Bull., 460.
RAGSDALE, A.C., THOMPSON, H.J., WORSTELL, D.M. and BRODY, S. (1951): Influences of increasing temperature, 40° to 105° F, on milk production in Brown Swiss cows, and on feed and water consumption and body weight in Brown Swiss and Brahman cows and heifers. Monogr. agr. exp. Stat. Res. Bull., 471.
RAGSDALE, A.C., THOMPSON, H.J., WORSTELL, D.M. and BRODY, S. (1953): Effect of humidity on milk production and composition, feed and water consumption, and body weight in cattle. Monogr. agr. exp. Stat. Res. Bull., 521.
REGAN, W.M. and RICHARDSON, G.A. (1938): Reactions of the dairy cow to changes in environmental temperature. J. Dairy Sci. 21: 73.
RICHARDSON, C.W. (1960): Effects of various combinations of environmental temperature and humidity on the composition of milk and milk fat of the dairy cow. Ph. D. Thesis, Univ. Missouri.
RICHARDSON, C.W. and JOHNSON, H.D. (1961): Effects of environmental temperature and humidity on the fatty acid composition of milk fat. J. Dairy Sci., 44: 1937-1940.
ROUSSEL, J.D., ORTEGO, J.D., GHOLSON, J.H. and FRYE, J.B., JR. (1969): Effect of thermal stress on the incidence of abnormal milk. J. Dairy Sci., 52: 912.
SARGENT, F.D., BUTCHER, K.R. and LEGATES, J.E. (1967): Environmental influences on milk constituents. J. Dairy Sci., 50: 177-184.
SEMENTOVSKAYA, R.M. and GARKAVI, O.V. (1950): This reference taken from: Milk, The Mammary Gland and Its Secretion. S.K. Kon and A.T. Cowie, Academic Press, New York.
WEATHER BUREAU (1959): Instructions to Field Stations for Experimental Use of „Discomfort Index." March.
WITZEL, S.A. and HEIZER, E.E. (1946): A 5-year summary of dairy barn research. Agr. Eng., 25: 499.
WORSTELL, D.M. and BRODY, S. (1953): Comparative physiological reactions of European and Indian cattle to changing temperature. Monogr. agr. exp. Stat. Res. Bull., 515.
WRIGHT, N.C. (1946): Report of the development of cattle breeding and milk production in Ceylon. Sessional Paper 20. Ceylon Govt. Press, Colombo.

Chapter 10, Section 1b: Meat Quality with Special Reference to the Effect of Heat on Beef Quality (D.F. DOWLING)

BERG, R.T. and BUTTERFIELD, R.M. (1966): Important considerations in beef carcass appraisal. Annual Feeders Day Report. Univ. Alberta, Edmonton, Canada, 45: 23-28.
BERG, R.T. (1969): Growth and carcass composition of heifers, steers and bulls. Annual Feeders Day Report. Univ. Alberta, Edmonton, Canada, 48: 24-28
BUTTERFIELD, R.M. (1963): A study of musculature of the steer carcass. Ph.D. Thesis, Univ. Queensland.
CHARLES, D.D. (1964): Classifying trade beef by specifications. Aust. vet. J., 40: 27-29.
DOWLING, D.F. (1955): The thickness of cattle skin. Aust. J. agric. Res., 6: 776–785.
DOWLING, D.F. (1956a): An experimental study of heat tolerance of cattle. Aust. J. agric. Res., 7: 469–481.
DOWLING, D.F. (1956b): Heat tolerance in animals. Aust. vet. J., 32: 246–249.
DOWLING, D.F. (1958): Seasonal changes in coat characters in cattle. Proc. Aust. Soc. anim. Prod., 2: 69–80.
DOWLING, D.F. (1960): In: Cattle Country. F.H. Johnston Publ. Co.: 147–154.
DOWLING, D.F. (1970): Improvement of beef cattle strains. Proc. Aust. Soc. anim. Prod., 8: 86–90.

GRIFFITH-DAVIES, J. and EYLES, A.G. (1965): Expansion of Australian pastoral production. J. Aust. Inst. agric. Sci., 31: 77–93.
HUTTON, E.M. (1970): In: Proc. Int. Grassland Congr., Surfers Paradise, 1970.
WEBSTER, A.J.F. (1969): In: Biometeorology, 4. S.W. Tromp and W.H. Weihe (eds) : 46.
WHALLEY, R.D.B. (1970): Exotic or native species – the orientation of pasture research in Australia. J. Aust. Inst. agric. Sci., 36: 111–118.
WHYTE, R.O. (1962): The myth of tropical grasslands. Trop. Agr. (Trin)., 39: 1–11.

Chapter 10, Section 2a: General Effects on Sheep (J.M. DONEY and J. SLEE)

ALEKSEEVA, G.I. (1962): Physiological characters of Merino-type sheep during acclimatization in Uzbekistan. In Russian. Zivotnovodstvo, 24, 9: 75-78.
ALEXANDER, G. (1962): Temperature regulation in the newborn lamb. V Summit metabolism. Aust. J. agr. Res., 13: 100-121.
ALEXANDER, G. (1964): Lamb survival: physiological considerations. In: Proc. Aust. Soc. anim. Prod., V: 113-122.
ALEXANDER, G. and WILLIAMS, D. (1966): Heat stress and growth of the conceptus in sheep. In: Proc. Aust. Soc. anim. Prod., 6: 102-105.
ALEXANDER, G. and WILLIAMS, D. (1971): Heat stress and development of the conceptus in domestic sheep. J. agr. Sci., 76: 53-72.
AREHART, L.A. JR, LEWIS, J.M., HINDS, F.C. and MANSFIELD, M.E. (1969): Space allowances for lambs on slatted floors. J. anim.Sci., 29: 638-641.
ARMSTRONG, D.G., BLAXTER, K.L., GRAHAM, N.McC. and WAINMAN, F.W. (1959): The effect of environmental conditions on food utilization by sheep. Anim. Prod. 1: 1-12.
BAIHANOV, M. (1965): The duration of oestrus and the sexual cycle in Yaidara ewes. Referat. Zh. Zhivot. Vet. 1967 (2) No. 2, 58, 126 (Abstr.).
BENNETT, J.W. and HUTCHINSON, J.C.D. (1964): Thermal insulation of short lengths of Merino fleece. Aust. J. agr. Res., 15: 427-445.
BLAXTER, K.L., GRAHAM, N.McC. and WAINMAN, F.W. (1959): Environmental temperature, energy metabolism and heat regulation in sheep. III. The metabolism and thermal exchanges of sheep with fleeces. J. agr. Sci., 52: 41-49.
BLAXTER, K.L. (1964a): Problems of cold environments. In: Proc. 6th int. Congr. Nutr : 227-242.
BLAXTER, K.L. (1964b): The effect of outdoor climate in Scotland on sheep and cattle. Vet. Rec., 76: 1445-1451.
BLAXTER, K.L., CLAPPERTON, J.L. and WAINMAN, F.W. (1966): The extent of differences between six British breeds of sheep in their metabolism, feed intake and utilization and resistance to climatic stress. Brit. J. Nutr., 20: 283-294.
BLIGH, J., INGRAM, D.L., KEYNES, R.D. and ROBINSON, S.G. (1965): The deep body temperature of an unrestrained Welsh Mountain sheep recorded by a radiotelemetric technique during a 12-month period. J. Physiol., 176: 136-144.
BLIGH, J. (1967): A thesis concerning the processes of secretion and discharge of sweat. Environm. Res., 1: 28-45.
BROCKWAY, J.M., MCDONALD, J.D. and PULLAR, J.D. (1965): Evaporative heat loss mechanisms in sheep. J. Physiol., 179: 554-568.
BROOKS, J.R. and ROSS, G.V. (1962): Effect of ambient temperature and thyroxine therapy on semen quality of rams. J. anim. Sci., 21: 700-705.
BURNS, MARCA (1968): Some observations on the adaptation of livestock to tropical environments. Ghana J. Sci., 9: 41-49.
COCKREM, F. (1967): The effect of temperature on body growth and other traits of open- and woolly-faced Romney lambs. In: Proc. N.Z. Soc. anim. Prod., 27: 193-209.
COCKREM, F.R.M. and MCDONALD, M.F. (1969): An investigation of the relationships between body temperature and implantation and lambing rates in the New Zealand Romney ewe. In: Proc. N.Z. Soc. anim. Prod., 29: 195-207.

CZERNEK, S., KLUZ, I. and JANKOWSKI, S. (1960): Comparison of the results of rearing long-woolled sheep in half-open and in closed sheep pens. Roczn. Nauk. rol., B, 76: 651-697.

DJANUAR, R. (1965): Effect of high temperature on spermiogenesis of rams. Communications vet., 9: 13-18.

DONEY, J.M. (1963): The effects of exposure in Blackface sheep with particular reference to the role of the fleece. J. agr. Sci., 60: 267-273.

DONEY, J.M. (1966): Breed differences in response of wool growth to annual nutritional and climate cycles. J. agr. Sci., 67: 25-30.

DONEY, J.M. and RUSSEL, A.J.F. (1968): Differences amongst breeds of sheep in food requirements for maintenance and live-weight change. J. agr. Sci., 71: 343-349.

DUTT, R.H. (1963): Critical period for early embryo mortality in ewes exposed to high ambient temperature. J. anim. Sci., 22: 713-719.

DUTT, R.H. (1964): Detrimental effects of high ambient temperature on fertility and early embryo survival in sheep. Int. J. Biometeor , 8: 47-56.

EDEY, T.N. (1969): Prenatal mortality in sheep: A review. Anim. Breed. Abstr. 37: 173-190.

ENTWISTLE, K.W. (1970): Fertilization rate and early embryonic mortality in ewes in north west Queensland. In: Proc. Aust. Soc. anim. Prod., 8:353-357.

EYAL, E. (1963): Shorn and unshorn Awassi sheep. I. Body temperature, II. Pulse rate, III. Respiration rate, IV. Skin temperatures and changes in temperature and humidity in the fleece and its surface. J. agr. Sci., 60: 159-168; 169-173; 175-181; 183-193.

FOWLER, D.G. and DUN, R.B. (1966): Skin folds and Merino breeding. 4. The susceptibility of rams selected for a high degree of skin wrinkle to heat-induced infertility. Aust. J. exp. Agr. anim. Husb., 6: 121-127.

FOWLER, D.G. (1968a): Skin folds and Merino breeding. 5. Variations in scrotal, testis and rectal temperature as affected by site of measurement, acclimatization to heat and degree of skin fold. Aust. J. exp. Agr. anim. Husb., 8: 125-132.

FOWLER, D.G. (1968b): Skin folds and Merino breeding. 7. The relations of heat applied to the testis and scrotal thermo-regulation to fertility in the Merino ram. Aust. J. exp. Agr anim. Husb., 8: 142-148.

FOWLER, D.G. and KENNEDY, J.P. (1968): Skin folds and Merino breeding 6. The effects of varying heat exposures and degree of skin fold on rectal, scrotal and testis temperatures. Aust. J. exp. Agr. anim. Husb., 8: 133-141.

GRAHAM, N. McC. (1964): Influence of ambient temperature on the heat production of pregnant ewes. Aust. J. agr. Res., 15: 982-988.

GRIFFITHS, J.G. and DONEY, J.M. (1969): Physiological adjustment to repeated wind exposure in sheep. Anim. Prod. 11: 493-498.

GRIFFITHS, J.G., GUNN, R.G. and DONEY, J.M. (1970): Fertility in Scottish Blackface ewes as influenced by climatic stress. J. agr. Sci., 75: 485-488.

HOWARTH, B. JR. (1969): Fertility in the ram following exposure to elevated ambient temperature and humidity. J. Reprod. Fertil., 19: 179-183.

HUTCHINSON, K.J. (1966): Bodyweight criteria and post-shearing deaths. In: Proc. Aust. Soc anim. Prod., 6: 229-233.

HUTCHINSON, J.C.D. (1968): Deaths of sheep after shearing. Aust. J. exp. Agr. anim. Husb., 8: 393-400.

JOYCE, J.P. and BLAXTER, K.L. (1964): The effect of air movement, air temperature and infrared radiation on the energy requirements of sheep. Brit. J. Nutr., 18: 5-27.

JOYCE, J.P., BLAXTER, K.L. and PARK, C. (1966): The effect of natural outdoor environments on the energy requirements of sheep. Res. vet. Sci., 7: 342-359.

KIJATKIN, P.F. and TAPILJSKII, I.A. (1959): Thermoregulation and metabolism in imported Merino and in indigenous fat-rumped rams. Dokl. Akad. Nauk. USSR, 12: 47-50.

KING, G. and CASSIE, A.B.D. (1940): Propagation of temperature changes through textiles in humid atmospheres. I. Rate of absorption of water vapour by wool fibres. Trans. Faraday Soc., 36: 445-453.

KURZ, V. (1955): Results of improvement of hill sheep in C.S.R. Sbornik V.S.Z.L. Brno, 2: 111-116.

LAMONT, J.L. (1964); Seasonal fluctuations in the semen quality of Merino rams in South Africa. In: Proc. S. Afr. Soc. anim. Prod., 1964, 3: 43-55.

LEES, J.L. (1966): Variations in the time of onset of the breeding season in Clun ewes. J. agr. Sci., 67: 173-179.

LINDSAY, D.R. (1969): Sexual activity and semen production of rams at high temperatures. J. Reprod. Fertil., 18: 1-8.

LOGGINS, P.E., KOGER, M., WARNICK, A.C. and CUNHA, T.J. (1964): Spring lamb production in Florida. Bull. Fla. agr. exp. Stat., 669: 20 pp.

LOKVANCIC, H. (1964): A contribution to the study of the annual sexual rhythm of domestic sheep with special regard to the effect of macroclimate factors. II. Some ovarian functions and macroclimate factors. Veterinaria (Sarafevo), 13: 99-106.

MACFARLANE, W.V. (1964): Terrestrial animals in dry heat: Ungulates. In: Handbook of Physiology, Sect. 4: Adaptation to the Environment. Amer. physiol. Soc., Washington D.C. 509-539.

MACFARLANE, W.V. (1969): Heat, cold and water in mammals. In: Biometeorology 4, S.W. Tromp and W.H. Weihe (eds.).

MCLAUGHLIN, J.W., EGAN, J.K., POYNTON, W.McL. and THOMPSON, R.L. (1970): The effect upon neonatal lamb motility of lambing systems incorporating the use of complete and partial shelter. In: Proc. Aust. Soc. anim. Prod., 8: 337-343.

MADISON, V.L. (1966): An experiment on the ultraviolet irradiation of lambs of the Romanov breed in winter. Referat. Zh. Zhivot. Vet. 1967, (7) No. 7 58.10.

MAKEVNIN, S.G. (1963): Physiological characters of sheep kept at high temperatures and in artificial shade during the summer heat. Referat. Zh. Zhivot. Vet. 1964 (7) No. 7.58.43.

MALIKOV, D.I. (1963): The effect of atmospheric pressure on reproduction in rams. Vestn. nel-hoz. Nauk (Mork) 8 (6): 82-86.

MARSHALL, T. and DIXON, F.F. (1969): Shed lambing in the Esperance environment. J. Agr. West. Aust., Ser. 4, 10: 513-514.

MIKUS, M. (1964): On the problems of open air (cold) rearing of sheep. Vedecke prace V.U.O. Trencin, Bratislava, S.V.P.L., 2: 133-151.

MILLER, G.R. (1968): Some responses of hill ewes and lambs to artificial shelter. Anim. Prod., 10: 59-66.

MOULE, G.R. and WAITES, G.M.H. (1963): Seminal degeneration in the ram and its relation to the temperature of the scrotum. J. Reprod. Fertil., 5: 433-446.

MOULE, G.R. (1966): Ovine reproduction in tropical Australia. Aust. vet. J., 42: 13-18.

MUNRO, J. (1962): The use of natural shelter by hill sheep. Anim. Prod., 4: 343-349.

NASYROV, Ju. Ja (1969): Acclimatization of Merinos and their crossbreds in Uzbekistan. Izv. kuibyshevsk. sel'-khoz. Inst. 25:193-196.

NEVILLE, W.E. JR. and NEATHERY, M.W. (1970): Effect of natural differences in atmospheric temperature on the incidence of first estrus following anestrus in ewes. J. anim. Sci., 30: 424-429.

OBST, J.M. and DAY, H.R. (1968): The effect of inclement weather on mortality of Merino and Corriedale lambs on Kangaroo Island. In: Proc. Aust. Soc. anim. Prod., 7: 239-242.

PANARETTO, B.A. (1968): Some metabolic effects of cold stress on undernourished non-pregnant ewes. Aust. J. agr. Res., 19: 273-282.

PANARETTO, B.A., HUTCHINSON, J.C.D. and BENNETT, J.W. (1968): Protection afforded by plastic coats to shorn sheep. In: Proc. Aust. Soc. anim. Prod., 7: 264-269.

PANARETTO, B.A. and FERGUSON, K.A. (1969): Pituitary-adrenal interactions in shorn sheep exposed to cold wet conditions. Aust. J. agr. Res., 20: 99-113.

PARER, J.T. (1963a): Wool length and radiant heating effects in sheep. J. agr. Sci., 60: 141.

PARER, J.T. (1963b): The effects of wool length and nutrition on heat reactions of Merino sheep in the field. Aust. J. exp. Agr. anim. Husb., 3: 243-248.

POPOVICI, S., STANCIU, G. and REBREANU, L. (1966): Changes in some biological characters in sheep during the adaptation to new environmental conditions. Lucr. stiint Inst. agron. Timisoara Ser. Med. vet., 9: 401-405.

PURSER, A.F. (1967): Animal Breeding Research Organisation Report: 26-34.

PURSER, A.F. and KARAM, H.A. (1967): Lamb survival, growth and fleece production in relation to birthcoat type among Welsh Mountain sheep. Anim. Prod., 9: 75-85.
QUARTERMAIN, A.R. (1964): Heat tolerance in Southern Rhodesian sheep fed on a maintenance diet. J. agr. Sci., 62: 333-339.
RATHORE, A.K. and YEATES, N.T.M. (1967): Morphological changes in ram spermatozoa due to heat stress. Vet. Rec., 81: 343-344.
RATHORE, A.K. (1968): Effects of high temperature on sperm morphology and subsequent fertility in Merino sheep. Proc. Aust. Soc. anim. Prod., 7: 270-274.
RATHORE, A.K. (1970): A note on the relation between scrotal wool cover and fertility in Merino rams. Anim. Prod., 12: 165-167.
REA, J.C., ROSS, C.V. and PFANDER, W.H. (1963): Studies of factors affecting performance of growing-finishing lambs. Res. Bull. M.O. agr. exp. Stat., 827: 65 pp.
RICH, T.D. and ALLISTON, C.W. (1970): The influence of programmed circadian temperature changes in the reproductive performance of ewes. J. anim. Sci., 30: 966-969.
RIEK, P.F., HARDY, M.H., LEE, D.H.K. and CARTER, H.B. (1950): The effect of the dietary plane upon the reactions of two breeds of sheep during short exposures to hot environments. Aust. J. agr. Res., 1: 217-230.
ROBINSON, W.I., BROWN, W. and LUCAS, I.A.M. (1970): Exposure studies on sheep at the University College of North Wales, Bangor. 3rd Symp. Shelter Res., Cambridge 1969. Min. of Agr.: 95-101.
RUSSEL, A.J.F., DONEY, J.M. and REID, R.L. (1967): Energy requirements of the pregnant ewe. J. agr. Sci., 68: 359-363.
RUSSEL, A.J.F. and DONEY, J.M. (1969): Observations on the use of plasma free fatty acid concentrations in the determination of maintenance requirements of sheep. J. agr. Sci., 72: 59-63.
RUSSEL, A.J.F. and DONEY, J.M. (1970): Measurement of the cost of climatic exposure of sheep. 3rd Symp. Shelter Res., MAFF. London 1970: 103-112.
RYLE, M. and MORRIS, L.R. (1961): Some quantitative studies on tissues of lambs dwarfed by high temperature during gestation. Aust. J. exp. Biol. med. Sci., 39: 79-91.
SAHNI, K.L. and ROY, A. (1967): A note on summer sterility in Romney Marsh rams under tropical conditions. Indian J. vet. Sci., 37: 335-338.
SETH, O.N., SARASWAT, K.C. and ROY, A. (1966): Respiration and rectal temperature as measures of adaptability of different breeds and crosses of lambs and kids to tropical environments. Agra Univ. J. Res. Sci., 15, 3: 17-30.
SHAFIE, N.M. and SHARAFELDIN (1965): Animal behaviour in the subtropics 1. Heat tolerance in relation to grazing behaviour in sheep. Neth. J. agr. Sci., 13: 1-5.
SHELTON, M. and HUSTON, J.E. (1968): Effects of high temperature stress during gestation on certain aspects of reproduction in the ewe. J. anim. Sci., 27: 153-158.
SLEE, J. (1966): Variation in the responses of shorn sheep to cold exposure. Anim. Prod., 8: 425-434.
SLEE, J. and SYKES, A.R. (1967): Acclimatization of Scottish Blackface sheep to cold. I. Rectal temperature responses. Anim. Prod., 9: 333-347.
SLEE, J. (1968): Body temperature and vasomotor responses in Scottish Blackface and Tasmanian Merino sheep subjected to slow cooling. Anim. Prod., 10: 265-282.
SLEE, J. (1970a): Resistance to body cooling in male and female sheep, and the effects of previous exposure to chronic cold, acute cold and repeated short cold shocks. Anim. Prod., 12: 13-21.
SLEE, J. (1970b): 3rd Symp. Shelter Res., Cambridge 1969: Min. of Agr.: 87-93.
SMITH, I.D. (1965): Reproductive wastage in Merino sheep in semi-arid tropical Queensland. Observations on flocks mated during the spring and summer. Aust. J. exp. Agr. anim. Husb., 5: 110-114.
SMITH, I.D., BELL, G.H. and DE CHANEET, G. (1966): Embryonic mortality in Merino ewes exposed to high ambient temperatures. Aust. vet. J., 42: 468-470.
SPIES, H.G., MENZIES, C.S., SCOTT, S.P., COON, L.L. and KIRACOFE, G.H. (1965): Effects of forced exercise and cooling on reproductive performance of finewool ewes bred during the summer. J. anim. Sci., 24: 9-12.

STEYNBERG, H. and HUGO, W.J. (1964): Effect of shade on the semen quality of rams. S. Afr. J. agr. Sci., 7: 101-107.

STOJANOV, A. (1967): The atmospheric temperature and humidity in various types of sheep houses and their effect on the development and productivity of the sheep. Zhivot Nauk., 4, 3: 73-84.

SYKES, A.R. and SLEE, J. (1968): Acclimatization of Scottish Blackface sheep to cold. 2. Skin temperature, heart rate, respiration rate, shivering intensity and skinfold thickness. Anim. Prod., 10: 17-35.

SYKES, A.R. and SLEE, J. (1969a): Cold exposure of Southdown and Welsh Mountain sheep. 1. Effects of breed, plane of nutrition and acclimatization to cold upon resistance to body cooling. Anim. Prod., 11: 65-75.

SYKES, A.R. and SLEE, J. (1969b): Cold exposure of Southdown and Welsh Mountain sheep. 2. Effects of breed, plane of nutrition and previous acclimatization to cold upon skin temperature, heart rate, shivering and respiration rate. Anim. Prod., 11: 77-89.

SYMINGTON, R.B. (1960): Studies on the adaptability of three breeds of sheep to a tropical environment. J. agr. Sci., 55: 287-300.

TANEJA, G.C. (1966): Physiological responses in sheep to arid environments. 1. Comparative response of indigenous and exotic breeds. Indian J. Physiol. all. Sci., 20: 24-29.

TE BRAKE, J.H.A. (1969): De seizoen variatie in geslachtsactiviteit bij de ovi; en het induceren van oestrus en ovulatie gedurende de anoestrusperiode. Lit. Overz. Cent. Landbouwpubl. Landbouw Docum., Wageningen, 35: 39 pp.

THWAITES, C.J. (1967a): Embryo mortality in the heat stressed ewe. I. The influence of breed. J. Reprod. Fertil., 14: 5-14.

THWAITES, C.J. (1967b): Age and heat tolerance in sheep. Int. J. Biometeor., 11: 209-212.

THWAITES, C.J. (1968): The influence of age of ewe on embryo mortality under heat stress conditions. Int. J. Biometeor., 12: 29-33.

THWAITES, C.J. (1969): Embryo mortality in the heat stressed ewe. II. Application of hot-room results to field conditions. J. Reprod. Fertil., 19: 255-262.

THWAITES, C.J. (1970): Embryo mortality in the heat stressed ewe. III. The role of the corpus luteum, thyroid and adrenal glands. J. Reprod. Fertil., 21: 95-107.

TYAGI, J.C. and LAVANIA, J.P. (1969): Breeding seasons of indigenous and cross-bred ewes and role of temperature and bright sunshine hours. Indian J. anim. Sci., 39: 91-94.

WAITES, G.M.H. and VOGLMAYR, J.K. (1963): The functional activity and control of the apocrine sweat glands of the scrotum of the ram. Aust. J. agr. Res., 14: 839-851.

WATSON, R.H., ALEXANDER, G., CUMMING, I.A., MACDONALD, J.W., MCLAUGHLIN, J.W., RIZZOLI, D.J. and WILLIAMS, D. (1968): Reduction of perinatal loss of lambs in winter in Western Victoria by lambing in sheltered individual pens. In: Proc. Aust. Soc. anim. Prod. 7: 243-249.

WEBSTER, A.J.F. and BLAXTER, K.L. (1966): The thermal regulation of 2 breeds of sheep exposed to air temperatures below freezing point. Res. vet. Sci., 7: 466-479.

WEBSTER, M.E.D. and LYNCH, J.J. (1966): Some physiological and behavioural consequences of shearing. In: Proc. Aust. Soc. anim. Prod., 6: 234-239.

WEBSTER, A.J.F. and PARK, C. (1967): The effect of jute coats on the heat losses of two breeds of sheep exposed to different environments. Anim. Prod., 9: 483-490.

WEBSTER, M.E.D. and JOHNSON, K.G. (1968): Some aspects of body temperature regulation in sheep. J. agr. Sci., 71: 61-66.

WEBSTER, A.J.F., HICKS, A.M. and HAYS, F.L. (1969): Cold climate and cold temperature induced changes in the heat production and thermal insulation of sheep. Canad. J. Physiol. Pharmacol., 47: 553-562.

WIENER, G. and SLEE, J. (1969): Relationships between inbreeding, plane of nutrition and temperature stress in sheep. In: Proc. 2nd Wld. Conf. anim. Prod., Maryland, U.S.A.: 490-491.

WINFIELD, C.J., BROWN, W. and LUCAS, I.A.M. (1968): Some effects of compulsory exposure over winter on in-lamb Welsh Mountain ewes. Anim. Prod., 10: 451-463.

WODZICKA, M. (1960): Seasonal variations in wool growth and heat tolerance of sheep. II. Heat tolerance. Aust. J. agr. Res., 11: 85-96.
WODZICKA-TOMASZEWSKA, M. (1963): The effect of shearing on the appetite of sheep. N.Z. J. agr. Res., 6: 440-447.
WOODY, C.O. and ULBERG, L.C. (1964): Viability of one-cell sheep ova as affected by high environmental temperature. J. Reprod. Fertil., 7: 275-280.
YEATES, N.T.M. (1958): Foetal dwarfism in sheep - an effect of high atmospheric temperature during gestation. J. agr. Sci., 51: 84-89.

Chapter 10, Section 2b: Effect on Woolproduction of Sheep (J.C.D. HUTCHINSON)

ALEXANDER, G. and WILLIAMS, D. (1971): Heat stress and development of the conceptus in domestic sheep. J. agr. Sci., 62: 59-66.
BENNETT, J.W., HUTCHINSON, J.C.D. and WODZICKA-TOMASZEWSKA, M. (1962): Annual rhythm of wool growth. Nature (Lond.), 194: 651-652.
BROWN, G.D. and WILLIAMS, O.B. (1970): Geographical distribution of the productivity of sheep in Australia. J. Aust. Inst. agr. Sci., 36: 182-198.
DONEY, J.M. (1964): The fleece of the Scottish Blackface sheep. IV. The effects of pregnancy, lactation and nutrition on seasonal wool production. J. agr. Sci., 62: 59-66.
DONEY, J.M. (1966): Breed difference in response of wool growth to annual nutritional and climatic cycles. J. agr. Sci., 67: 25-30.
DONEY, J.M. and EADIE, J. (1967a): The wool growth response to the annual cycle of grazing intake in Cheviot wethers. J. agr. Sci., 69: 411-416.
DONEY, J.M. and GRIFFITHS, J.G. (1967b): Wool growth regulation by local skin cooling. Anim. Prod., 9: 393-397.
DOWNES, A.M. and HUTCHINSON, J.C.D. (1969): Effect of low skin surface temperature on wool growth. J. agr. Sci., 72: 155-158.
DUNLOP, A.A. and HAYMAN, R.H. (1958): Differences among Merino strains in resistance to fleece-rot. Aust. J. agr. Res., 9: 260-266.
GEYTENBEEK, P.E. (1963): A survey of post-shearing losses due to adverse weather conditions. Expl. Rec. Dept. Agr. S. Aust., 1: 21-30.
HAYMAN, R.H. (1953): Studies in fleece-rot of sheep. Aust. J. agr. Res., 4: 430-468.
HAYMAN, R.H. (1955): Studies in fleece-rot of sheep: Some ecological aspects. Aust. J. agr. Res., 6: 466-475.
HUTCHINSON, J.C.D. and WODZICKA-TOMASZEWSKA, M. (1961): Climate physiology of sheep. Anim. Breed. Abstr., 29: 1-14.
HUTCHINSON, J.C.D. (1965): Photoperiodic control of the annual rhythm of wool growth. In: Biology of the Skin and Hair Growth. A.G. Lyne and B.F. Short (eds). Angus & Robertson, Sydney: 565-573.
HUTCHINSON, J.C.D. (1968): Deaths of sheep after shearing. Aust. J. exp. Agr. anim. Husb., 8: 393-400.
HUTCHINSON, J.C.D. and BENNETT, J.W. (1969): Design and use of plastic coats to protect shorn sheep. C.S.I.R.O. Div. anim. Physiol. Techn. Rep. no. 1/69.
HUTCHINSON, J.C.D. and BENNETT, J.W. (1970): Further tests of plastic coats for shorn sheep. C.S.I.R.O. Div. anim. Physiol. Techn. Rep. no. 2/70.
HUTCHINSON, J.C.D., BENNETT, J.W., O'CALLAGHAN, M. and JONES, P.N. (In preparation a): Skin surface temperature and wool growth: effect of a cyclical temperature treatment and of prolonged exposure to a low skin temperature. J. agr. Sci.
HUTCHINSON, J.C.D., BENNETT, J.W., WODZICKA-TOMASZEWSKA, M., O'CALLAGHAN, M. and JONES, P.N. (In preparation b): Effect of ambient temperature and daylength on the annual rhythm of wool growth in Merino and Southdown ewes. J. agr. Sci.
HUTCHINSON, K.J. (1962): Climate corrections to the seasonal wool growth rhythm of sheep grazing in a southern Australian environment. In: Proc. Aust. Soc. anim. Prod., 4: 34-45.
JOLLY, M. and LYNE, A.G. (1970): The response of skin and wool growth to local subdermal temperature changes in sheep. J. agr. Sci., 75: 501-515.

LEEPER, G.W. (ed.) (1970): The Australian Environment. Commonwlth sci. industr. Res. Org. Australia, in association with Melbourne Univ. Press, Melbourne.
LIPSON, M., ELLINGWORTH, J.S. and SINCLAIR, J.F. (1970): Effect of sheeprugging on fleece properties and processing of wool. Textile Inst. and Industry, 8: 100-102.
LYNE, A.G., JOLLY, M. and HOLLIS, D.E. (1970): Effects of experimentally produced local subdermal temperature changes on skin temperature and wool growth in the sheep. J. agr. Sci., 74: 83-90.
NAGARCENKAR, R. and BHATTACHARYA, P. (1964a): Effect of environmental factors on wool growth. 1. Effect of normal climatic factors. Indian J. vet. Sci., 34: 89-99.
NAGARCENKAR, R. and BHATTACHARYA, P. (1964b): Effect of environmental factors on wool growth. II. Effects of photoperiodicity. Indian J. vet. Sci., 34: 100-107.
NAGARCENKAR, R. and BHATTACHARYA, P. (1964c): Factors responsible for canary colouration of the wool. Indian J. vet. Sci., 34: 46-60.
NAGARCENKAR, R. and BHATTACHARYA, P. (1964d): Relationship of certain pelt characteristics with „ canary colouration" of wool. Indian J. vet. Sci., 34: 242-252.
NARAYAN, S. and SHARMA, R.S. (1968): Some observations on fleece characters, wool yellowing and skin glands output in Nali sheep located in different states. Wool and Woollens of India, 5: 33-37.
NARAYAN, S., CLARKE, W.H. and SHARMA, R.S. (1969): Resistance to yellow fleece stain in Sonadi sheep. Indian vet. J., 46: 865-872.
PANARETTO, B.A., HUTCHINSON, J.C.D. and BENNETT, J.W. (1968): Protection afforded by plastic coats to shorn sheep. In: Proc. Aust. Soc. anim. Prod., 7: 264-269.
PENGELLEY, E.T. (1967): The relation of external conditions to the onset and termination of hibernation and estivation. In: Mammalian Hibernation, 3. Oliver & Boyd, Edinburgh: 1-29.
ROBINSON, D.W. (1969): Preliminary observations on the heat tolerance of shorn and nutritionally depleted sheep in a tropical environment. Brit. vet. J., 125: 112-120.
ROUGEOT, J. (1957): Action de la variation saisonnière de la durée quotidienne d'éclairement sur la mue de certaines fibres de la toison de la race ovine Limousine. C.R. Soc. Biol. (Paris), 151: 834-837.
ROUGEOT, J. (1961): Actions comparées des variations périodiques, annuelles et semestrielles, de la durée quotidienne de l'éclairement sur les cycles des follicules des jarres courts de la toison des brebis Limousines. Relations avec leur cycle de reproduction. Ann. Biol. anim., 1: 385-402.
SLEE, J. (1965): Seasonal patterns of moulting in Wiltshire Horn sheep. In: Biology of the Skin and Hair Growth. A.G. Lyne and B.F. Short (eds). Angus & Robertson, Sydney: 543-563.
THWAITES, C.J. (1967): Prolonged heat stress and wool growth in sheep. Int. J. Biometeor., 11: 297-300.
THWAITES, C.J. (1968): Heat stress and wool growth in sheep. In: Proc. Aust. Soc. anim. Prod., 7: 259-263.
WODZICKA-TOMASZEWSKA, M. and BIGHAM, M.L. (1968): The tattooed patch technique for measurement of wool growth. N.Z. J. agr. Res., 11: 943-947.

Chapter 10, Section 3: Effects on Pigs (D.L. INGRAM)

ADAM, TH. (1967): Der Einfluss des Stallklimas auf einige Fuktionen der Ferkel und Mastschweine. Mh. vet.–med., 22: 752-756.
BALDWIN, B.A. and INGRAM, D.L. (1967): Behavioural thermoregulation in pigs. Physiol. and Behav. 2: 15-21.
BALDWIN, B.A. and INGRAM, D.L. (1968a): Factors influencing behavioural thermoregulation in pigs. Physiol. and Behav., 3: 409-415.
BALDWIN, B.A. and INGRAM, D.L. (1968b): The role of behaviour in the regulation of body temperature. Vet. Ann., 9: 130-142.

BOAZ, T.G. and WALTERS, A.A. (1966): Weight and growth rate. Pig Farming, 14: 45-47.

BOND, T.E., KELLY, C.F. and HEITMAN, H. (1963): Effect of diurnal temperature on heat loss and well-being of swine. Trans. Amer. Soc. agr. Eng., 6: 132-135.

BOND, T.E., HEITMAN, H. and KELLY, C.F. (1965): Effects of increased air velocities on heat and moisture loss and growth of swine. Trans. Amer. Soc. agr. Eng., 8: 167-174.

BOND, T.E., KELLY, C.F. and HEITMAN, H. (1967): Physiological response of swine to cycling environmental conditions. Anim. Prod., 9: 453-462.

CAIRNIE, A.B. and PULLAR, J.D. (1959): An investigation into the efficient use of time in calorimetric measurement of heat output. Brit. J. Nutr., 13: 431-439.

CUNNINGHAM, J.M.M. and ROWELL, J.G. (1963): Is insulation over-rated? Pig Farming, March: 41-43.

CURTIS, S.E. (1970): Environmental-thermoregulatory interactions and neonatal piglet survival. J. anim. Sci., 31: 576-587.

FULLER, M.F. (1965): The effect of environmental temperature on the nitrogen metabolism and growth of the young pig. Brit. J. Nutr., 19: 531-546.

GORDON, W.A.M. (1962): Environmental studies in pig housing. I. Air Velocity. Brit. vet. J., 118: 171-205.

GUNNARSON, H.J., BUTCHBAKER, A.F., WITZ, R.L. and DINUSSON, W.E. (1967): Effect of air velocity, air temperature and mean radiant temperature on performance of growing finishing swine. Trans. Amer. Soc. agr. Eng., 10: 715-722.

HARMAN, D.J., DALE, A.C. and JONES, H.W. (1968): Effect of floor type on required moisture-vapour removal rate from swine finishing houses. Trans. Amer. Soc. agr. Eng., 11: 149-152.

HEAL, J.W., INGRAM, D.L. and LEGGE, K.F. (1970): Measurement of heat loss from the skin by means of radiotelemetry. J. Physiol. (Lond.), 210: 123P.

HEITMAN, H., KELLY, C.F. and BOND, T.E. (1958): Ambient air temperature and weight gain in swine. J. anim. Sci., 17: 62-67.

HEITMAN, H., HAHN, L., BOND, T.E. and KELLY, C.F. (1962): The effects of modified summer environment on swine behaviour. Anim. Behav., 10: 15-19.

HERBST, H.G. (1967): Der postnatale Temperaturverlauf beim Schwein und einige seiner Bedingungen. Mh. vet.—med., 22: 731-735.

HOLME, D.W. and COEY, W.E. (1967): The effects of environmental temperature and method of feeding on the performance and carcass composition of bacon pigs. Anim. Prod., 9: 209-218.

HOLMES, C.W. and MOUNT, L.E. (1967): Heat loss from groups of growing pigs under various conditions of environmental temperature and air movement. Anim. Prod., 9: 435-452.

HOLUB, A. (1967): Temperatur-regulation bei Ferkeln im ersten Monat ihres Lebens. Mh. vet.—med., 22: 726-731.

INGRAM, D.L. (1964a): The effect of environmental temperature on body temperature, respiratory frequency and pulse rate in the young pig. Res. vet. Sci., 5: 348-356.

INGRAM, D.L. (1964b): The effect of environmental temperature on heat loss and thermal insulation in the young pig. Res. vet. Sci., 5: 357-364.

INGRAM, D.L. (1965): The effect of humidity on temperature regulation and cutaneous water loss in the young pig. Res. vet. Sci., 6: 9-17.

INGRAM, D.L. (1965): Evaporative cooling in the pig. Nature (Lond.), 207: 415-416.

INGRAM, D.L. and MOUNT, L.E. (1965): Metabolic rate of young pigs living at high ambient temperatures. Res. vet. Sci., 6: 300-306.

INGRAM, D.L. and SLEBODZINSKI, A. (1965): Oxygen consumption and thyroid gland activity during adaptation to high ambient temperatures in young pigs. Res. vet. Sci., 6: 522-530.

INGRAM, D.L. (1967): Stimulation of cutaneous glands in the pig. J. comp. Path., 77: 93-98.

INGRAM, D.L. and WEAVER, M.E. (1969): A quantitative study of blood vessels of the pig's skin and the influence of environmental temperature. Anat. Rec., 163: 517-524.

INGRAM, D.L. and LEGGE, K.F. (1970): The thermoregulatory behaviour of young pigs in a natural environment. Physiol. and Behav., 5: 981-987.

INGRAM, D.L. and LEGGE, K.F. (1970): The effect of environmental temperature on respiratory ventilation in the pig. Resp. Physiol., 8: 1-12.
INGRAM, D.L. and LEGGE, K.F. (1970): Variations in deep body temperature in the young unrestrained pig over the 24–hr period. J. Physiol. (Lond.), 210: 989-998.
IRVING, L. (1964): Terrestrial animals in cold: birds and mammals. In: Adaptation to Environment. D.B. Dill (ed.). Amer. physiol. Soc.: 361-377.
JENSEN, A.H., KAHLMAN, D.E., BECKER, D.E. and HARMON, B.G. (1969): Response of growing-finishing swine to different housing environments during winter seasons. J. anim. Sci., 29: 451-456.
KELLY, C.F., BOND, T.E. and GARRETT, W. (1964): Heat transfer from swine to a cold slab. Trans. Amer. Soc. agr. Eng., 7: 34-37.
LEUCHT, W. (1967): Zur Untersuchungsmethodik der Thermoregulation beim neugeborenen Schwein. Mh. vet.– med., 22: 735-739.
LYHS, L. (1967): Grundlagen und neuere Erkenntnisse auf dem Gebiet der Temperaturregulation. Mh. vet. – Med., 22: 821-726.
LYHS, L., NICHELMANN, M. and STEINHARDT, M. (1969): Hauttemperatur des Schweines bei verschiedener Umgebungstemperatur. Mh. vet.–med., 24: 494-496.
MACGRATH, W.S., VANDER NOOT, G.W., GILBREATH, R.L. and FISHER, H. (1968): Influence of environmental temperature and dietary fat on back fat composition in swine. J. Nutr., 96: 461-466.
MACHARDY, F.W. (1960): Transient heat flow in piggery floors. Canad. agr. Eng., 2: 13-15.
MORRISON, S.R., BOND, T.E. and HEITMAN, H. (1967): Skin and lung moisture loss from swine. Trans. Amer. Soc. agr. Eng., 10: 691-696.
MORRISON, S.R., BOND. T.E. and HEITMAN, H. (1968): Effect of humidity on swine at high temperatures. Trans. Amer. Soc. agr. Eng., 11: 526-528.
MOUNT, L.E. (1960): The influence of huddling and body size on the metabolic rate of the young pig. J. agr. Sci., 55: 101-105.
MOUNT, L.E. (1962): Evaporative heat loss in the new-born pig. J. Physiol. (Lond.), 164: 274-281.
MOUNT, L.E. (1964): Radiant and convective heat loss from the new-born pig. J. Physiol. (Lond.), 173: 96-113.
MOUNT, L.E. and INGRAM, D.L. (1964): The effect of ambient temperature and air movement on localized heat loss from the pig. Res. vet. Sci., 6: 84-91.
MOUNT, L.E. (1966): The effect of wind-speed on heat production in the new-born pig. Quart. J. exp. Physiol., 51: 18-26.
MOUNT, L.E. (1967): The heat loss from new-born pigs to the floor. Res. vet. Sci., 8: 175-186.
MOUNT, L.E. (1968): The Climatic Physiology of the Pig. Ed. Arnold, London.
NICHELMANN, M. (1967): Temperaturauswahlvermögen neugeborener Ferkel. Mh. vet.– med., 22: 739-744.
PFEIFFER, H. (1967): Einfluss der Umgebungstemperatur und Luftfeuchtigkeit auf die Leistungsfähigkeit von Mastschweinen. Mh. vet.– med., 22: 980-985.
ROLLER, W.L. and GOLDMAN, R.F. (1969): Response of swine to acute heat exposure. Trans. Amer. Soc. agr. Eng., 12: 164-169.
SCHÜLKE, B., BEEGER, E., GÜTHE, B. and KADING, O. (1967): Dynamik einiger Puffersysteme beim Schwein unter hohen Umgebungstemperaturen. Mh. Vet.– med., 22: 749-752.
SEMAN, J. (1967): The effect of outdoor weather on the micro-climate in stables. I. The effect of outdoor temperature and humidity on the air temperature and humidity in farrowing pigstyes. Acta Univ. agr. Brno, B., Fac. vet., 36: 168-178.
SEMAN, J. (1968): The effect of outdoor weather on the micro-climate in stables. II. The effect of winds on the air temperature and humidity in farrowing pigstyes. Acta Univ. agr. Brno, B., Fac. vet., 37: 339-346.
SEYMOUR, E.W., SPEER, V.C., HAYS, V.W., MANGOLD, D.W. and HAZEN, T.E. (1964): The effects of dietary protein level and environmental temperature on performance and carcass quality on growing finishing swine. J. anim. Sci., 23: 375-379.

SØRENSEN, P.H. (1962): Influence of climatic environment on pig performance. In: Nutrition of Pigs and Poultry, J.T. Morgan and D. Lewis (eds). Butterworths, London: 88-103.
STEPHENS, D.B. and START, I.B. (1970): The effects of ambient temperature, nature and temperature of the floor and radiant heat on the metabolic rate of the new-born pig. Int. J. Biometeor., 14: 275-283.
STEPHENS, D.B. (1971): The metabolic rates of new-born pigs in relation to floor insulation and ambient temperatures. Anim. Prod., in press.
STUDZINSKI, T. and STEINHARDT, M. (1967): Besonderheiten und Grenzen der Anionen-Kationen-Regulation beim Schwein bei der Thermoregulation. Mh. Vet.– med., 22: 744-749.
TEAGUE, H.S., ROLLER, W.L. and GRIFO, A.P. (1968): Influence of high temperatures on reproductive performance in swine. J. anim. Sci., 27: 408-411.
TONKS, H.M. and SMITH, W.C. (1965): The effect of a high temperature, high humidity indoor environment on the performance of bacon pigs. Anim. Prod., 7: 280.
WEAVER, M.E. and INGRAM, D.L. (1969): Morphological changes in swine associated with environmental temperature. Ecology, 50: 710-713.

Chapter 10, Section 4: Effects on Poultry, with Special Reference to Disease, Egg Production and Quality (W.O. WILSON)

ACKERS, T.G., BOND, J., GOLDBERG, L.J. (1966): Effect of temperature and relative humidity on survival of airborne Columbia SK group virus. Appl. Microbiol., 14: 361-366.
ANDERSON, D.P., BEARD, C.W. and HANSON, R.P. (1964): The adverse effects of ammonia on chickens including resistance to infection with Newcastle disease virus. Avian Dis., 8: 369-379.
ANDERSON, D.P., BEARD, C.W. and HANSON, P.P. (1966): Influence of poultry house dust, ammonia and carbon dioxide on the resistance of chickens to Newcastle Disease virus. Avian Dis., 10: 178-188.
ANDERSON, D.P., WOLFE, R.R., CHERMS, F.L. and ROPER, W.E. (1968): Influence of dust and ammonia on the development of air sac lesions in turkeys. Amer. J. vet. Res., 29: 1049-1058.
BORNSTEIN and LEV, Y. (1969): Studies with decreasing lighting regimes for early-season replacement pullets under subtropical conditions. 1. A comparison of various programs. Poultry Sci., 48: 1919-1933. II. A possible explanation for their effect on egg size. Poultry Sci., 48: 1934-1939.
BRAY, D.J. and GESSELL, J.A. (1961): Studies on corn soya laying diets. 4. Environmental temperature. Poultry Sci., 40: 1328-1335.
BRAY, D.J. (1968): Photoperiodism and age as factors affecting the protein requirements of laying pullets. Poultry Sci., 47: 1005-1013.
CHARLES, D.R. (1970): Poultry environment in the U.K. A review of progress. Wld Poultry Sci., 26: 422-434.
DEATON, J.W., REECE, F.N. and BOUCHILLON, C.W. (1969): Heat and moisture production of broilers. 2. Winter conditions. Poultry Sci., 48: 1579-1582.
DRURY, L.N., PATTERSON, W.C. and BEARD, C.W. (1969): Ventilating poultry houses with filtered air under positive pressure to prevent airborne diseases. Poultry Sci., 48: 1640-1646.
EL-BOUSHY, A.R., SIMONS, P.C.M. and WIERTZ, G. (1968): Structure and ultrastructure of the hen's egg shell as influenced by environmental temperature, humidity and vitamin C additions. Poultry Sci., 47: 456-467.
EL HALAWANI, M. El. S., WILSON, W.O. and BURGER, R.E. (1970): Cold-acclimation and the role of catecholamines in body temperature regulation in male Leghorns. Poultry Sci., 49: 621-632.
GRIFFIN, J.G. and VARDAMAN, T.H. (1970): Diurnal cyclic versus daily constant temperatures for broiler performances. Poultry Sci., 49: 387-392.

GRIFFIN, J.G., MCNALLY, E.H. and VARDAMAN, T.H. (1967): The effects of ventilation rates on broiler condemnation and performance. In: 64th Ann. Proc. Ass. agr. Workers: 313-314.

HARRISON, P.C., BIELLIER, H.V. and FUNK, E.M. (1965): Effect of rate of decrease and increase in photoperiod on sexual maturity of S. C. White Leghorn pullets. Poultry Sci., 44: 1377.

HOWES, J.R. (1968): Tools for measuring environment, temperature, humidity/air movement, illumination, atmospheric composition, bird density. Feedstuffs, 40, 28: 14.

LACASSAGNE, L. and JACQUET, J.D. (1965): Maturité sexuelle et apparition d'oeufs à double jaune et sans coquille. Rôle du régime lumineux et du niveau alimentaire. Ann. Zootechn., 14: 169-179.

LEIGHTON, A.T. JR. and POTTER, L.M. (1969): Reproductive performance of turkeys subjected to blackout versus brownout restricted light conditions. Poultr Sci., 48: 505-514.

LILLIE, R.J. and DENTON, C.A. (1965): Effect of lighting systems in the grower and adult periods upon the over-all performance of White Leghorns. Poultry Sci., 44: 809-816.

LILLIE, R.J. (1969): A literature review of air pollutants affecting poultry. Poultry Sci., 48: 1835.

LILLIE, R.J. (1970): Air pollutants affecting the performance of domestic animals. A literature review. U.S.D.A., ARS agr. Handbook, 380. Washington, D.C.

LONG, J., DALE, A.C., HARINGTON, R.B. and JOHNSON, R.C. (1966): The influence of selected temperatures and ventilation rates on egg production. Poultry Sci., 45: 58-65.

LINSLEY, J.G. and BURGER, R.E. (1964): Respiratory and cardiovascular responses in the hyperthermic domestic cock. Poultry Sci., 43: 291-305.

MORRIS, T.R. (1966): Light requirements of the fowl. Symp. „Environmental control in poultry production." Brit. Egg Market. Bd.: 1-35.

MORRIS, T.R. and OWEN, V.M. (1966): The effect of light intensity on egg production. In: Proc. 13th Wld Poultry Sci. Congr.: 458-461.

MORRIS, T.R. and FOX, S. (1968): The response of pullets of various ages to different amounts and qualities of light. Edit. S. Fox, Farm Bldg Centre.

NOLES, R.K. and SMITH, R.E. (1964): Rearing and maintaining pullets on controlled lighting. Poultry Sci., 43: 848-857.

PARKER, J.E. and McCLUSKEY, W.H. (1964): The effect of the length of daily light periods on the volume and fertilizing capacity of semen from male chickens. Poultry Sci., 43: 1401-1405.

PARKER, J.E. and McCLUSKEY, W.H. (1965): The effect of length of daily light periods on sexual development and subsequent fertilizing capacity of male chickens. Poultry Sci., 44: 23-27.

PAYNE, C.G., LINCOLN, D.W. and CHARLES, D.R. (1965): The influence of constant and fluctuating environmental temperature on time of oviposition under continuous lighting. Brit. Poultry Sci., 6: 93-95.

PAYNE, C.G. (1966): Practical aspects of environmental temperature for laying hens. Wld Poultry Sci. J., 22: 126-139.

REECE, F.N., DEATON, J.W. and VARDAMAN, T.H. (1969): The effect of temperature exposure on broiler condemnations. Poultry Sci., 48: 288-294.

REID, B.L., KURNICK, A.A. and HULETT, A.A. (1965): Relationship of protein level, age and ambient temperature to laying hen performance. Poultry Sci., 44: 1113-1122.

SHUTZE, J.V. (1966): Recent developments in lighting programs. Feedstuffs, 38, 14: 26-29.

SIEGEL, H.S. and DRURY, L.N. (1968): Physiological responses to high lethal temperature and air velocity in young fowl. Poultry Sci., 41: 1230-1235.

SIEGEL, H.S. (1969): Environmental stress and physiological compensating mechanism in fowl-temperature and respiratory regulation. Poultry Sci., 48: 22-30.

SIEGEL, H.S. and DRURY, L.N. (1970): Broiler growth in diurnal cycling temperature environments. Poultry Sci., 49: 238-244.

SUMMERS, J.D. (1965): Temperature and diet on egg production and quality. Canad. Poultry Rev., 89: 24-32.
TANAKA, K., MATHER, F.B., WILSON, W.O., MCFARLAND, L.Z. (1965): Effects of photoperiods on early growth of gonads and on potency of gonadotropins of the anterior pituitary in coturnix. Poultry Sci., 44: 662-665.
VALENTINE, H. (1964): A study of the effect of different ventilation rates on the ammonia concentrations in the atmospheric of broiler houses. Brit. Poultry Sci., 5: 149-159.
WALTER, C.W. (1966): Cross infection in hospitals. A.S.H.R.A.E. J., 8: 41-49.
WHITTOW, G.C. (1965): Regulation of Body Temperature. In: Avian Physiology, 2nd Ed. P.D. Sturkie (ed.).: 186-238.
WILSON, W.O., WOODARD, A.E. and ABPLANALP, H. (1964): Exogenous regulation of oviposition in chickens. Poultry Sci., 43: 1187-1192.
WILSON, W.O., MATHER, F.B., BENNETT, E. and WOODARD, A.E. (1964): Association between blood spots in eggs and ambient temperature. Poultry Sci., 43: 1607-1698.
WILSON, W.O., OGASAWARA, F.X. and WOODARD, A.E. (1967): Increasing egg production in winter hatched turkeys by a restricted light treatment. Poultry Sci., 46: 46-52.
WILSON, W.O. and SIOPES, T.D. (1967): Conditioned gonadal response of Japanese quail to cyclic photoperiods. Int. J. Biometeor., 11: 289-295.
WINN, P.N. and GODFREY, E.F. (1967): The effect of humidity on growth and feed conversion of broiler chickens. Int. J. Biometeor., 11: 39-50.
WITTER, R.L. (1970): Epidemiological studies related to the control of Marek's disease. Wld Poultry Sci. J., 26: 755-762.
WOLFORD, J.H. and TANAKA, K. (1970): Factors influencing egg shell quality. A review. Wld Poultry Sci., 26: 763-780.
YAMADA, Y., ITO, S. and ISHIDA, E. (1966): The effect of monthly hatches on performance traits in White Leghorn pullets. Jap. Poultry Sci., 3: 181-191.

Chapter 10, Section 5: Fertility and Reproduction (D. SMIDT)

ALLISTON, C.W., HOWARTH, B. JR. and ULBERG, L.C. (1965): Embryonic mortality following culture in vitro of one- and two-cell rabbit eggs at elevated temperatures. J. Reprod. Fertil., 9: 337-341.
DJANUAR, R. (1965): Effect of high temperature on spermiogenesis of rams. Commun. vet. 9: 13-18.
DUTT, R.H. (1960): Temperature and light as factors in reproduction among farm animals. J. Dairy Sci., 43: 123-139.
DUTT, R.H. (1964): Detrimental effects of high ambient temperature on fertility and early embryo survival in sheep. Int. J. Biometeor. 8: 47-56.
FARNER, D.S. and FOLLET, B.K. (1966): Light and other environmental factors affecting avian reproduction. J. anim. Sci., 25, Suppl.: 90-117.
HAFEZ, E.S.E. (1964): Effects of high temperature on reproduction. Int. J. Biometeor. 7: 223-230.
HAFEZ, E.S.E. (1965): Environment and reproduction in farm animals. Wld Rev. anim. Prod., 1: 1-11.
HAFEZ, E.S.E. (1968): Environmental effects on animal productivity. In: Adaptation of Domestic Animals. E.S.E. Hafez (ed.). Lea & Febiger Philadelphia: 74-97.
JÖCHLE, W. (1966): The circadian rhythm in female reproduction. In: Reproduction in the Female Mammal. G.E. Lamming and F.C. Amoroso (eds). Butterworths, London: 267-285.
SHELTON, M. (1964): Relationship of environment temperature during gestation to birth weight and mortality of lambs. J. anim. Sci., 23: 360-364.
SMITH, I.D., BELL, G.H. and DE CHANEET, G. (1966): Embryonic mortality in Merino ewes exposed to high ambient temperatures. Aust. vet. J., 42: 468-470.

STOTT, G.H. and WILLIAMS, R.J. (1962): Causes of low breeding efficiency in dairy cattle associated with seasonal high temperature. J. Dairy Sci., 45: 1369-1375.

THIBAULT, C., COUROT, M., MARTINET, L., MAULEON, P., DUMESNILL, DUBOISSON, F., ORTAVANT, R., PELLETIER, J. and SIGNORET, J.P. (1966): Regulation of breeding season and estrons cycles by light and external stimuli in some mammals. J. anim. Sci., 25, Suppl.: 119-139.

TOMPKINS, E.C., HEIDENREICH, C.J. and STOB, M. (1967): Effect of postbreeding thermal stress on embryonic mortality in swine. J. anim. Sci., 26: 377-380.

ULBERG, L.C. (1967): Effects of macro- and microenvironment of the biology of mammalian reproduction. In: Ground Level Climatology, R.H. Shaw (ed.), Amer. Ass. Adv. Sci. Publ. No. 86, Washington D.C.

ULBERG, L.C. (1969): Embryonaler Frühtod als spezielles fortpflanzungsbiologisches Problem an tropischen Standorten. Tierzüchter, 20: 168-171.

WAITES, G.M.H. (1968): Temperature and Fertility in Mammals. VI Congr. Int. Reprod. anim. Insem. Artif., Paris. Vol I: 235-245.

YEATES, N.T.M. (1958): Foetal dwarfism in sheep, an effect of high atmospheric temperature during gestation. J. agr. Sci., 51: 84-89.

YEATES, N.T.M. and PARER, J.T. (1962): Effects of stress on reproduction. In: Reproduction in Farm Animals, E.S.E. Hafez (ed.). Lea & Febiger, Philadelphia: 297-303.

Chapter 10, Section 6: Animal Disease (W. FERGUSON AND D. BRANAGAN)

ALEXANDER, A.F., WILL, D.H., GROVER, R.F. and REEVES, T.T. (1960): Pulmonary hypertension and right ventricular hypertrophy in cattle at high altitudes. Amer. J. vet. Res., 21: 199-204.

ANDERSON, D.E. and SKINNER, P.E. (1961): Studies on bovine ocular squamous carcinoma („cancer eye"). XI. Effects of sunlight. J. anim. Sci., 20: 474-477.

ANDREWS, C.H. (1964): The complex epidemiology of respiratory virus infections. Science, 146: 1274-1277.

ARTHUR, D.R. (1962): Ticks and disease. Pergamon Press, Oxford/London/New York/Paris.

BIERER, B.W. (1962): Artificially induced Escherichia coli infection in chilled chicks. Poultry Sci., 41: 1627 (abstract).

BRANAGAN, D. and HAMMOND, J.A. (1965): Rinderpest in Tanganyika: a review. Bull. epizoot. Dis. Afr., 13: 225-246.

BRANAGAN, D. (1970): Survival and development of Rhipicephalus appendiculatus Neumann, 1901 in laboratory and field. Ph. D. Thesis, Univ. Edinburgh.

BUXTON, P.A. (1955): The natural history of tsetse flies. Mem. No. 10, London Sch. Hyg. and trop. Med. H.K. Lewis, London.

CARE, A.D. (1967): Absorption and hypomagnesaemia. Feed Forum, 2: 6-9.

COYLE, T.J. (1961): The epidemiology of Fasciola gigantica in Uganda Protectorate. Fellowsh. Thesis, Roy. Coll. vet. Surg., London.

DALHAM, T. (1956): Mucous flow and ciliary activity in the trachea of healthy rats and rats exposed to repiratory irritant gases. Acta physiol. Scand., 36, suppl. 123, Stockholm.

DAVIES, S.F.M., JOYNER, L.P. and KENDALL, S.B. (1963): Coccidiosis. Oliver and Boyd, Edinburgh/London.

DOGIEL, V.A. (1964): General parasitology. Oliver and Boyd, Edinburgh/London.

EKMEIER, H. and AGDE, K. (1965): Trockene Stalluft als Krankheitsursache bei Schweinen. Dtsch. tierärztl. Wschr., 72: 385-386.

FOTHERINGHAM, W. and LEWIS, E.A. (1937): East Coast fever: its transmission by ticks in Kenya Colony. Hyalomma impressum near planum P. sch. as a vector. Parasitology, 29: 504-523.

FREEDEN, H.T. (1965): Genetic aspects of disease resistance. Anim. Breed. Abstr., 33: 17-26.

GREGSON, J.D. (1951): As quoted by Arthur (1962).

HAMMOND, J.A. (1965): Observations on fascioliasis in Tanganyika. Bull. epizoot. Dis. Afr., 13: 55-65.

HARRISON, P.C., LAUBER, J.K. and MCGINNIS, J. (1969). Light-induced eye abnormalities in domestic fowl. In: Biometeorology 4. Proc. 5th Int. Biom. Congr. S.W. Tromp and W.H. Weihe (eds), II:98-99.

HUGH JONES, M.E. and AUSTWICK, P.K.C. (1967). Epidemiological studies in bovine mycotic abortion. I. The effect of climate on incidence. Vet. Rec., 81:273-276.

HURST, G.W. (1968). Foot and mouth disease. The possibility of continental sources of the virus in England in the epidemics of October, 1967, and several other years. Vet. Rec., 82; 610-614.

HYSLOP, N.S.G. (1969): Viability of infective organisms in moving airstreams. In: Biometeorology 4. Proc. 5th Int. Biomet. Congr., S.W. Tromp and W.H. Weihe (eds.), 1: 198-199.

JEBSON, J.L. (1950): Salmonellosis in sheep. Aust. vet. J., 26: 256-258.

JORDAN, A.M. (1962): The ecology of the fusca group of tsetse flies (*Glossina*) in southern Nigeria. Bull. ent. Res., 53: 355-385.

JORDAN, A.M. (1963): The distribution of the fusca group of tsetse flies (*Glossina*) in Nigeria and West Cameroun. Bull. ent. Res., 54: 307-323.

KHEISIN, E.M. and LAVRENENKO, L.E. (1956): Duration of blood-sucking and daily rhythm of feeding and dropping in females of *Ixodes ricinus* L. (in Russian). Zool. Zh., 35: 379-383.

LOVAT EVANS, C., NISBET, A.M. and ROSS, K.A. (1957): A histological study of the sweat glands of normal and dry-coated horses. J. comp. Path., 67: 397-405.

LUTTA, A.S. and KHEISIN, E.M. (1954): As quoted by Arthur (1962).

MACLEOD, J. (1932): Preliminary studies in the tick transmission of louping ill. Vet. J., 88: 276-284.

MACLEOD, J. (1939): The seasonal and annual incidence of the sheep tick *Ixodes ricinus* L. in Britain. Bull. ent. Res., 30: 103-118.

MILNE, A. (1946): The ecology of *Ixodes ricinus* L. - distribution on hill pasture. Parasitology, 37: 75-81.

MILNE, A. (1947): The ecology of *Ixodes ricinus* L. - differential host levels on hill pastures. Parasitology, 38: 27-50.

MILNE, A. (1948): The ecology of *Ixodes ricinus*. - review of hosts. Parasitology, 39: 167-197.

MINETT, F.C. (1952): The annual and seasonal incidence of anthrax in various countries. Climatic effects and sources of infection. Bull. Off. int. Epizoot., 37: 238-300.

NASH, T.A.M. (1939): The ecology of the puparium of Glossina in Northern Nigeria. Bull. ent. Res., 30: 259-284.

OLLERENSHAW, C.B. (1966): The approach to forecasting the incidence of fascioliasis over England and Wales, 1958 - 1962. Agr. Met., 3: 35-53 (Centr. vet. Lab., Weybridge).

OLLERENSHAW, C.B. and SMITH, L.P. (1966): An empirical approach to forecasting the incidence of nematodiriasis over England and Wales. Vet. Rec., 79: 536-540.

OLLERENSHAW, C.B. and SMITH, L.P. (1969): A relationship between climate and the incidence of swayback in England and Wales, and its application to forecasting the incidence of disease. Proc. 5th Int. Biometeor. Congr., S.W. Tromp and W.H. Weihe (eds), II: 100.

PAGE, W.A. (1959): Some observations on the fusca group of tsetse flies (*Glossina*) in the south of Nigeria. Bull. ent. Res., 50: 633-646.

PASTEUR, L., JOUBERT and CHAMBERLAND, C. (1878): Sur le charbon des poules. C.R. Acad. Sci., 87: 47-48.

PAVLOVSKI, E.N. (1947): Micro-organisms, their vectors and their external environment (in Russian). Zool. Zh., 26: 297-312.

ROBERTS, J.A. (1964): Enhancement of virulence of attenuated *Ectromelia* virus in mice maintained in a cold environment. Aust. J. exp. Biol. med. Sci., 42: 657-666.

ROBERTSHAW, D. and KATONGOLE, C.B. (1969): Adreno-cortical activity and intermediary metabolism of *Bos indicus* and *Bos taurus* in the high altitude (2,000 m) tropics. In: Biometeorology 4. Proc. 5th Int. Biometeor. Congr. S.W. Tromp and W.H. Weihe (eds), II: 101.

SAINSBURY, D.W.B. (1968): Environment, housing and disease. Poultry Rev. (May and Baker Publ.), 8: 30-40.

SCHOLES, J.C. and HUTT, F.B. (1942): The relationship between body temperature and genetic resistance to *Salmonella pullorum* in the fowl. Mem. No. 244, Cornell Univ. Agr. exp. Stat., Ithaca, N.Y.

SELYE, H. (1936): A syndrome produced by diverse nocuous agents. Nature (Lond.), 138: 32.

SELYE, H. (1950): „Stress". Acta Inc., Montreal, Canada.

SILVERMAN, P.H. and CAMPBELL, J.A. (1959). Studies on parasitic worms of sheep in Scotland. I. Embryonic and larval development of *Haemonchus contortus* at constant conditions. Parasitology, 49: 23-38.

SINHA, S.K., HANSON, R.P. and BRANDLY, C.A. (1957): Effect of environmental temperature upon facility of aerosol transmission of infection and severity of Newcastle disease among chickens. J. infect. Dis., 100: 162-168.

SMITH, L.P. and HUGH JONES, M.E. (1969): The weather factor in foot and mouth epidemics. Nature (Lond.), 223: 712 715.

STAMPA, S. and du TOIT, R. (1958): Paralysis of stock due to the Karroo paralysis tick (*Ixodes rubicundus* Neu.). S. Afr. J. Sci., 54: 241-246.

STOTT, G.H. (1969): Adrenal involvement in reproductive failure in bovines related to high environmental temperature. In: Biometeorology 4. Proc. 5th Int. Biometeor. Congr. S.W. Tromp and W.H. Weihe (eds), II: 101.

TRIPP, J. (1965): The management of poultry disease from the advisory point of view. Poultry Rev. (May and Baker Publ.), 8: 30-40.

TROMP, S.W. (1968): The effect of local and regional meteorological environments on the development and spreading of communicable diseases (A Review). Rep. Biometeor. Res. Centre, Leiden, 11.

UNDERWOOD, E.J. (1962): Trace elements. Academic Press, New York/London.

WATTS, P.S. and WALL, M. (1952): The 1951 *Salmonella typhi-murium* epidemic in sheep in South Australia. Aust. vet. J., 28: 165-168.

WOODHEAD, T. (1968): a. Studies of potential evaporation in Kenya. b. Studies of potential evaporation in Tanzania. E. Afr. Agr. Forestry Res. Org., Box 30148, Nairobi, Kenya.

YEOMAN, G.H. (1966): Field vector studies of epizootic East Coast fever. Bull. epizoot. Dis. Afr., 14: 29-35 and 113-140.

YEOMAN, G.H. (1967): Field vector studies of epizootic East Coast fever. Bull. epizoot. Dis. Afr., 15: 89-113 and 189-212.

Chapter 10, Section 7: Arthropods (Including Insects, Ticks and Mites) as Vectors of Human and Animal Diseases (W.O. HAUFE)

ARAGÃO, M.B. (1964): Distribuiçao geográfica e abundância das espécies de *Anopheles* (Kerteszia) (Diptera, Culicidae). Revta bras. Malar., 16: 73-109.

BAST, T.F. (1961): A note on the flight periodicity of mosquitoes. Proc. New Jers. Mosq. Exterm. Ass., 48: 203-211.

BELOZEROV, V.N., BOGDANOV, V.E. and KVITKO, N.V. (1966): Seasonal changes in the temperature reaction of engorged females of the tick *Ixodes ricinus* L. (Ixodidae). (In Russian.) Vestn. leningr. gos. Univ., 21 (Ser. Biologia, Part 1): 37-44.

BELOZEROV, V.N. (1967): The water content and the regulation of water balance in females of Ixodid ticks (Acarina, Ixodidae) during feeding and after engorgement. (In Russian.) Zool. Zh., 46: 1182-1187.

CUMMING, J.E. and WRIGHT, R.H. (1967): Convection currents that guide mosquitoes. Canad. Ent., 99: 556-558.

DEPNER, K.R. (1962): Effects of photoperiod and of ultraviolet radiation on the incidence of diapause in the horn fly, *Haematobia irritans* (L.) (Diptera: Muscidae). Int. J. Biometeor., 5: 68-71.

DERGACHEVA, T.I. (1967): On the adaptation of sandflies (Diptera, Phlebotominae) to varying conditions of air humidity in colonies of the large gerbil and in villages in the Kashdadar'ya region of the Uzbek SSR. (In Russian.) Zool. Zh., 46: 1112-1116.

DOW, R.P. and GERRISH, G.M. (1970): Day-to-day change in relative humidity and the activity of *Culex nigripalpus* (Diptera: Culicidae). Ann. ent. Soc. Amer., 63: 995-999.

DYK, V. and BOUČKOVÁ, I. (1968): Die Temperatur-Feuchte-Relation in der Aktivität des gemeinen Holzbocks. Angew. Parasit., 9: 83-87.

ELDRIDGE, B.F. (1963): The influence of daily photoperiod on bloodfeeding activity of *Culex tritaeniorhynchus* Giles. Amer. J. Hyg., 77: 49-53.

FOSTER, R. (1964): Contributions to the epidemiology of human sleeping sickness in Liberia: bionomics of the vector *Glossina palpalis* (R.-D.) in a forest habitat. Bull. ent. Res., 54: 727-744.

FREDEEN, F.J.H. (1963): Oviposition in relation to the accumulation of blood-thirsty flies *Simulium (Gnus) arcticum* Mall. (Diptera) prior to a damaging outbreak. Nature (Lond.), 200: 1024.

GARRETT-JONES, C. (1962): The possibility of active long-distance migrations by *Anopheles pharoensis* Theobald. Bull. Wld Hlth Org., 27: 299-302.

GARRETT-JONES, C. (1964): Prognosis for interruptions of malaria transmission through assessment of the mosquito's vectorial capacity. Nature (Lond.), 204: 1173-1175.

GRAHAM, S.A. (1956): Ecology of forest insects. A. Rev. Ent., 1: 261-280.

GUIDICELLI, J. (1966): Récoltes de simulies en Côte-d'Ivoire. Étude de l'activité diurne des femelles de *Simulium damnosum* Theobald. Ann. Soc. ent. Fr., 2: 325-342.

HADDOW, A.J. (1964): Observations on the biting habits of mosquitoes in the forest canopy at Zika, Uganda, with special reference to the crepuscular periods. Bull. ent. Res., 55: 589-608.

HADDOW, A.J., CASLEY, D.J.L., O'SULLIVAN, J.P., ARDOIN, P.M.L., SSENKUBUGE, Y. and KITAMA, A. (1968): Entomological studies from a high steel tower in Zika Forest, Uganda. Part II. The biting activity of mosquitoes above the forest canopy in the hour after sunset. Trans. roy. ent. Soc. Lond., 120: 219-236.

HAINE, E. (1961): Do atmospheric electrical factors have an influence on the activity changes of small insects, in particular on the moulting and reproduction rates of aphids? Forsch-Ber. Landes NRhein-Westf., 974: 1-80.

HAINE, E. (1967): New trapping and recording methods for the study of migrations of virus-transmitting insects. Mitt. biol. BundAnst., Berlin-Dahlem, 121: 113-117.

HARWOOD, R.F. (1964): Physiological factors associated with male swarming of the mosquito *Culex tarsalis* Coq. Mosquito News, 24: 320-325.

HAUFE, W.O. (1957): Importance of air movement as a factor in mathematical expressions for biological response to environment. In: Proc. 1st Bioclimatol. Congr., Vienna. Int. J. Bioclimatol. Biometeor., I, III B.

HAUFE, W.O. (1961): Effect of adaptation on stimulation of flight in *Aedes aegypti* (L.) in a changing environment. Canad. Ent., 93: 522-539.

HAUFE, W.O. (1962): Response of *Aedes aegypti* (L.) to graded light stimuli. Canad. J. Zool., 40: 54-64.

HAUFE, W.O. (1963a): Insects as vectors of disease. In: Medical Biometeorology - Weather, Climate and the Living Organism. S.W. Tromp (ed.), Elsevier Publ. Co., Amsterdam: 688-703.

HAUFE, W.O. (1963b): Entomological biometeorology. Int. J. Biometeor., 7: 129-136.

HAUFE, W.O. (1963c): Ethological and statistical aspects of a quantal response in mosquitoes to environmental stimuli. Behaviour, 20: 221-241.

HAUFE, W.O. (1964a): Insects affecting the health of man. In: A Survey of Human Biometeorology. F. Sargent II and S.W. Tromp (ed.), Techn. Note No. 65, Wld met. Org., Geneva, pp. 90-95.

HAUFE, W.O. (1964b): Quantitative measurement of activity of *Aedes aegypti* (L.) (Culicidae: Diptera) in response to changes in the hygrothermal environment. Int. J. Biometeor., 7: 245-264.

HAUFE, W.O. (1966a): Synoptic correlation of weather with mosquito activity. In: Biometeorology II. Proc. 3rd Int. Biometeor. Congr., S.W. Tromp and W.H. Weihe (eds.). Pergamon Press, London: 523-540.

HAUFE, W.O. (1966b): The significance of biometeorology in the ecology of insects. Int. J. Biometeor., 10: 241-252.

HAYES, R.O. and HESS, A.O. (1964): Climatological conditions associated with outbreaks of eastern encephalitis. Amer. J. trop. Med. Hyg., 13: 851-858.

HELSON, G.A.H. and PENMAN, J.E.R. (1970): Biological transformations associated with weather changes. An hypothesis on flight activity of *Wiseana* spp. Int. J. Biometeor., 14: 227-246.

HUMPHRIES, D.A. (1967): Uptake of atmospheric water by the hen flea *Ceratophyllus gallinae* (Schrank). Nature (Lond.), 214-426.

HURST, G.W. (1969): Meteorological aspects of insect migrations. Endeavour, 28: 77-81.

JOHNSON, C.G. (1954): Aphid migration in relation to weather. Biol. Rev., 29: 87-118.

JOHNSON, C.G. (1969): Migration and dispersal of insects by flight. Methuen and Co. Ltd., London, 763 pp.

KETTLE, D.S. (1969): The biting habits of *Culicoides furens* (Poey) and *C. barbosai* Wirth and Blanten. II. Effect of meteorological conditions. Bull. ent. Res., 59: 241-258.

KNÜLLE, W. (1966): Equilibrium humidities and survival of some tick larvae. J. med. Ent., 2: 335-338.

KÜHLHORN, F. (1963): Populationsdynamische Untersuchungen bei *Anopheles messeae* Fall und *Anopheles claviger* Meig. (Dipt. Culicidae) in Oberbayern. (Untersuchungen über die Insektenfauna von Räumen: 9). Z. angew. Zool., 50: 311-327.

LOAEZA, R. MAC G. (1961): The distribution of mosquitoes in a humidity gradient. (In Spanish.) Acta Zool. Mex., 5: 1-9.

MAW, M.G. (1965): Effects of air ions in duration and rate of sustained flight of the blowfly, *Phaenicia sericata* Meigen. Canad. Ent., 97: 552-556.

MEOLA, R. (1964): The influence of temperature and humidity on embryonic longevity in *Aedes aegypti*. Ann. ent. Soc. Amer., 57: 468-472.

MESSENGER, P.S. (1959): Bioclimatic studies with insects. A. Rev. Ent., 4: 183-206.

MORRIS, K.R.S. (1962): The epidemiology of sleeping sickness in East Africa. V. Epidemics on the Albert Nile. Trans. roy. Soc. trop. Med. Hyg., 56: 316-338.

MULLA, M.S. and CHAUDHURY, M.F.B. (1968): Influence of some environmental factors on the viability and hatching of *Aedes aegypti* (L.) eggs. Mosquito News, 28: 217-221.

NANDA, D.K., SINGH, M.V. and CHAND DIWAN. (1962): Study of the effect of climate on the density of *C. fagitans* and the development of the filarial parasite in it. Indian J. Malar., 16: 313-320.

NIELSON, E.T. (1961): Twilight and the „crep" unit. Nature (Lond.), 190: 878-879.

NORRIS, K.R. (1966): Daily patterns of flight activity of blowflies (Calliphoridae: Diptera) in the Canberra district as indicated by trap catches. Aust. J. Zool., 14: 835-853.

OMER, S.M. and CLOUDSLEY-THOMPSON, J.L. (1968): Dry season biology of *Anopheles gambiae* Giles in the Sudan. Nature (Lond.), 217: 879-880.

PALIMPSESTOV, M.A. and GONCHAROV, A.P. (1960): Survival of eggs of *Psoroptes ovis* under various ecological conditions. (In Russian.) Sb. Trud. khar'kov. vet. Inst., 24: 315-320.

POWER, R.J.B. (1964): The activity pattern of *Glossina longipennis* Corti (Diptera: Muscidae). Proc. roy. ent. Soc. Lond. (A), 39: 5-14.

REUBEN, R. (1963): A comparison of trap catches of *Culicoides impunctatus* Goetghelves (Diptera: Ceratopogonidae) with meteorological data. In: Proc. roy. ent. Soc. Lond. (A), 38: 181-193.

RICHTER, C.J.J. (1970): Aerial dispersal in relation to habitat in eight wolf spider species (Pardosa, Araneae, Lycosidae). Oecologia (Berl.), 5: 200-214.

ROBINSON, N. and DIRNFELD, F.S. (1963): The ionization state of the atmosphere as a function of the meteorological elements of various sources of ions. Int. J. Biometeor., 6: 101-110.

ROWLEY, W.A. and GRAHAM, C.L. (1968): The effect of age on the flight performance of female *Aedes aegypti* mosquitoes. J. Insect Physiol., 14: 719-728.
ROWLEY, W.A. (1970): Interval flights and glycogen utilization by the mosquito, *Culex tarsalis*. J. Insect Physiol. 16: 1839-1844.
UVAROV, B.P. (1931): Insects and climate. Trans. roy. ent. Soc. Lond., 79: 1-247.
VARMA, M.G.R. (1965): The distribution of *Ixodes ricinus* in Britain in relation to climate and vegetation.B. Rosický and K.H. Heyberger, (ed.) Theoretical questions of natural foci of diseases. Publ. House Czech. Acad. Sci. (Prague): 301-311.
WELLINGTON, W.G. (1957): The synoptic approach to studies of insects and climate. A. Rev. Ent., 2: 143-162.
WHARTON, R.H. (1963): Equilibrium humidity. In: Advances in Acarology J.A. Naegle (ed.), Cornell Univ. Press, 1: 201-208.
WILLIAMS, C.B. (1958): Insect Migration. Collins, London: 235.
WILLIAMS, C.B. (1961): Studies in the effect of weather conditions on the activity and abundance of insect populations. Phil. Trans. B, 244: 331-378.
WINSTON, P.W. (1967): Cuticular water pump in insects. Nature (Lond.), 214: 383-384.
WRIGHT, R.H. and KELLOGG, F.E. (1962): Response of *Aedes aegypti* to moist convection currents. Nature (Lond.), 194: 402-403.

Chapter 10, Section 8: Laboratory Animals(W.H. WEIHE)

ANDERSON, A., WERBOFF, J. and LES, E.P. (1968): Effects of environmental temperature-humidity and cage density on body weight and behaviour in mice. Experientia (Basel), 24: 1022-1023.
BARNETT, S.A. (1959): The skin and hair of mice living at a low environmental temperature. Quart. J. exp. Physiol., 44: 35-42.
BARNETT, S.A., COLEMAN, E.M. and MANLY, B.M. (1959): Oxygen consumption and body fat of mice living at -3°C. Quart. J. exp. Physiol., 44: 43-51.
BARNETT, S.A. (1965): Adaptation of mice to cold. Biol. Rev., 40: 5-51.
BREWER, N.R. (1964): Estimating heat produced by laboratory animals. Heat. Pip. Air condit., 36(10): 139-141.
CABANAC, C. (1972): Thermoregulatory behavior. In: Essays on Temperature Regulation. J. Bligh and R. Moore (eds). Nort-Holland Publ. Co., Amsterdam: 19-36.
DE KONING, J., VAN DER WAAY, D., VOSSEN, J.M., VERSPRILLE, A. and DOOREN, L.J. (1970): Barrier nursing of an infant in a laminar cross-flow bench. Ned. Tijdschr. Kindergeneesk., 38: 1-13.
GESELLSCHAFT FÜR VERSUCHSTIERKUNDE (1971): Planung und Errichtung von Versuchstierbereichen tierexperimentell tätiger Institutionen. Juris-Druck-Verlag, Zürich: 1-45.
HARLAND, W.A. and BARNETT, S.A. (1969): Effects of sex, strain and environmental temperature on the composition of body lipids in mice. Quart. J. exp. Physiol., 54: 202-206.
HARRISON, G.H. and CLEGG, E.J. (1969): Environmental factors influencing mammalian growth. In: Physiology and Pathology of Adaptation Mechanisms. E. Bajusz (ed.), Pergamon Press, Oxford: 74-85.
HARRISON, G.A., MORTON, R.J. and WEINER, J.S. (1960): The growth in weight and tail length in inbred and hybrid mice reared at two different temperatures. Phil. Trans. B, 242: 479-516.
HARRISON, G.A., HIORNS, R.W. and WEINER, J.S. (1964): The growth of mice on fluctuating temperature environment. Proc. roy. Soc. B., 160: 137-148.
KLEIBER, M. (1961): The Fire of Life. An Introduction to Animal Energetics. Wiley & Sons, New York/London, 454 pp.
KNUDSON, B. (1962): Growth and reproduction of house mice at three different temperatures. Oikos, 13: 1-14.
LANE-PETTER, W. (ed.) (1963): Animals for Research. Academic Press London.

LEVENSON, S.M., DOFT, F., LEV, M. and KAN, D. (1969): Influence of micro-organisms on oxygen consumption. Carbon dioxide production and colonic temperature of rats. J. Nutr., 97: 542-562.
MOORE, K. (1944): The effect of controlled temperature changes on the behavior of the white rat. J. exp. Physiol. 34: 70-79.
MOUNT, L.E. (1971): Metabolic rate and thermal insulation in albino and hairless mice. J. Physiol. (Lond.), 217: 315-326.
MURAKAMI, H. (1971): Differences between internal and external environments of the mouse cage. Lab. Anim. Sci., 21: 680-684.
OLGILVIE, D.M. (1970): Behavioral thermoregulation and cold adaptation in two strains of white mice. Int. J. Biometeor., 14: 285-292.
PENNYCUIK, P.R. (1964): The effect on rats of chronic exposure to 34°C. IV Reproduction. Aust. J. biol. Sci., 17: 245-260.
PENNYCUIK, P.R. (1967a): A comparison of the effects of a variety of factors on the metabolic rate of the mouse. Aust. J. exp. Biol. med. Sci., 45: 331-346.
PENNYCUIK, P.R. (1967b): A comparison of the affects of a range of high environmental temperature and of two different periods of acclimatization on the reproductive performances of male and female mice. Aust. J. exp. Biol. med. Sci., 45: 527-532.
PENNYCUIK, P.R. (1969): Reproductive performance and body weights of mice maintained for 12 generations at 34°C. Aust. J. biol. Sci., 22: 667-675.
PRYCHODKO, W. (1958): Effect of aggregation of laboratory mice (Mus musculus) on food intake at different temperature. Ecology, 39: 500-503.
ROUBICEK, C.B., RAY, D.E. and TUCKER, H. (1968): Performance of rats at constant and intermittant temperatures and with four feeding schedules. Growth, 32: 135-141.
SCHOLANDER, P.F., HOCK, R., WALTERS, V. and IRVING, L. (1950): Adaptation to cold in arctic and tropical mammals and birds in relation to body temperature insulation and basal metabolic rate. Biol. Bull., 99: 259-271.
SIMMONS, M.L., ROBIE, D.M., JONES, J.B. and SERRANO, L.J. (1968): Effect of a filter cover on temperature and humidity in a mouse cage. Lab. Anim., 2: 113-120.
STILLE, G., BREZOWSKY, H. and WEIHE, W.H. (1968): The influence of the weather on the locomotor activity of mice. Arzneimittelforsch., 18: 892-893.
TOLCSVAI-NAGY, L., BREZOWSKY, H., STILLE, G. and WEIHE, W.H. (1968): The influence of the weather on the behaviour of cats. Arzneimittelforsch., 18: 1061-1063.
WEIHE, W.H., BREZOWSKY, H. and SCHWARZENBACH, F.R. (1961): Der Nachweis einer Wirkung des Klimas und Wetters auf das Wachstum der Ratte. Pflügers Arch. ges. Physiol., 273: 514-527.
WEIHE, W.H. (1964): Der Einfluss der Temperatur als Faktor der physikalischen Umwelt auf den Stoffwechsel der Ratte. In: Die Umwelt der Versuchstiere. W.H. Weihe (ed.). Int Z. Vitamin-forsch., 9: 89-113.
WEIHE, W.H. (1965a): Temperature and humidity climatograms for mice and rats. Lab. Anim. Care, 15: 18-28.
WEIHE, W.H. (1965b): Influence of altitude and cold on pregnancy and lactation of rats fed on two different diets. Int. J. Biometeor., 9: 43-52.
WEIHE, W.H. (1966): Das Körpergewicht der Ratte als Parameter zur Bestimmung der Höhenadaptation. Experientia (Basel), 22: 697-698.
WEIHE, W.H. (1970): Physiological responses to heat and cold in laboratory animals. In: The Effects of Environment. Carworth Europe, Collected Papers, Alconbury, Huntingdon, England, 4: 10-22.
WEIHE, W.H. (1971a): The significance of the physical environment for the health and state of adaptation of laboratory animals. In: Defining the Laboratory Animal. Nat. Acad. Sci., Washington D.C., 353-375.
WEIHE, W.H. (1971b): Effects of air velocity on hairless mice. J. Physiol. (Paris), 63: 452-454.
WEIHE, W.H. (1971c): Behavioral thermoregulation in single and grouped mice with change of cooling power of the air. Int. J. Biometeor., 15: 356-361.
WEIHE, W.H. (1972): The effect of temperature on the action of drugs. In: The Pharmacology of Thermoregulation. P. Lomax and E. Schönbaum (eds). Karger, Basel.

AUTHOR INDEX

SUBJECT INDEX